Climate: A Very Short Introduction

VERY SHORT INTRODUCTIONS are for anyone wanting a stimulating and accessible way in to a new subject. They are written by experts, and have been published in more than 25 languages worldwide.

The series began in 1995, and now represents a wide variety of topics in history, philosophy, religion, science, and the humanities. The VSI Library now contains more than 300 volumes—a Very Short Introduction to everything from ancient Egypt and Indian philosophy to conceptual art and cosmology—and will continue to grow in a variety of disciplines.

Very Short Introductions available now:

ADVERTISING Winston Fletcher
AFRICAN HISTORY John Parker and Richard Rathbone
AGNOSTICISM Robin Le Poidevin
AMERICAN HISTORY Paul S. Boyer
AMERICAN IMMIGRATION David A. Gerber
AMERICAN POLITICAL PARTIES AND ELECTIONS L. Sandy Maisel
AMERICAN POLITICS Richard M. Valelly
THE AMERICAN PRESIDENCY Charles O. Jones
ANAESTHESIA Aidan O'Donnell
ANARCHISM Colin Ward
ANCIENT EGYPT Ian Shaw
ANCIENT GREECE Paul Cartledge
ANCIENT PHILOSOPHY Julia Annas
ANCIENT WARFARE Harry Sidebottom
ANGELS David Albert Jones
ANGLICANISM Mark Chapman
THE ANGLO-SAXON AGE John Blair
THE ANIMAL KINGDOM Peter Holland
ANIMAL RIGHTS David DeGrazia
THE ANTARCTIC Klaus Dodds
ANTISEMITISM Steven Beller
ANXIETY Daniel Freeman and Jason Freeman
THE APOCRYPHAL GOSPELS Paul Foster
ARCHAEOLOGY Paul Bahn
ARCHITECTURE Andrew Ballantyne
ARISTOCRACY William Doyle
ARISTOTLE Jonathan Barnes
ART HISTORY Dana Arnold
ART THEORY Cynthia Freeland
ATHEISM Julian Baggini
AUGUSTINE Henry Chadwick
AUSTRALIA Kenneth Morgan
AUTISM Uta Frith
THE AVANT GARDE David Cottington
THE AZTECS Davíd Carrasco
BACTERIA Sebastian G. B. Amyes
BARTHES Jonathan Culler
BEAUTY Roger Scruton
BESTSELLERS John Sutherland
THE BIBLE John Riches
BIBLICAL ARCHAEOLOGY Eric H. Cline
BIOGRAPHY Hermione Lee
THE BLUES Elijah Wald
THE BOOK OF MORMON Terryl Givens
BORDERS Alexander C. Diener and Joshua Hagen
THE BRAIN Michael O'Shea
THE BRITISH CONSTITUTION Martin Loughlin
THE BRITISH EMPIRE Ashley Jackson
BRITISH POLITICS Anthony Wright
BUDDHA Michael Carrithers
BUDDHISM Damien Keown
BUDDHIST ETHICS Damien Keown
CANCER Nicholas James
CAPITALISM James Fulcher
CATHOLICISM Gerald O'Collins

THE CELL Terence Allen and Graham Cowling
THE CELTS Barry Cunliffe
CHAOS Leonard Smith
CHILDREN'S LITERATURE Kimberley Reynolds
CHINESE LITERATURE Sabina Knight
CHOICE THEORY Michael Allingham
CHRISTIAN ART Beth Williamson
CHRISTIAN ETHICS D. Stephen Long
CHRISTIANITY Linda Woodhead
CITIZENSHIP Richard Bellamy
CIVIL ENGINEERING David Muir Wood
CLASSICAL MYTHOLOGY Helen Morales
CLASSICS Mary Beard and John Henderson
CLAUSEWITZ Michael Howard
CLIMATE Mark Maslin
THE COLD WAR Robert McMahon
COLONIAL AMERICA Alan Taylor
COLONIAL LATIN AMERICAN LITERATURE Rolena Adorno
COMEDY Matthew Bevis
COMMUNISM Leslie Holmes
THE COMPUTER Darrel Ince
THE CONQUISTADORS Matthew Restall and Felipe Fernández-Armesto
CONSCIENCE Paul Strohm
CONSCIOUSNESS Susan Blackmore
CONTEMPORARY ART Julian Stallabrass
CONTEMPORARY FICTION Robert Eaglestone
CONTINENTAL PHILOSOPHY Simon Critchley
COSMOLOGY Peter Coles
CRITICAL THEORY Stephen Eric Bronner
THE CRUSADES Christopher Tyerman
CRYPTOGRAPHY Fred Piper and Sean Murphy
THE CULTURAL REVOLUTION Richard Curt Kraus
DADA AND SURREALISM David Hopkins
DARWIN Jonathan Howard
THE DEAD SEA SCROLLS Timothy Lim
DEMOCRACY Bernard Crick
DERRIDA Simon Glendinning
DESCARTES Tom Sorell
DESERTS Nick Middleton
DESIGN John Heskett
DEVELOPMENTAL BIOLOGY Lewis Wolpert
THE DEVIL Darren Oldridge
DICTIONARIES Lynda Mugglestone
DINOSAURS David Norman
DIPLOMACY Joseph M. Siracusa
DOCUMENTARY FILM Patricia Aufderheide
DREAMING J. Allan Hobson
DRUGS Leslie Iversen
DRUIDS Barry Cunliffe
EARLY MUSIC Thomas Forrest Kelly
THE EARTH Martin Redfern
ECONOMICS Partha Dasgupta
EDUCATION Gary Thomas
EGYPTIAN MYTH Geraldine Pinch
EIGHTEENTH-CENTURY BRITAIN Paul Langford
THE ELEMENTS Philip Ball
EMOTION Dylan Evans
EMPIRE Stephen Howe
ENGELS Terrell Carver
ENGINEERING David Blockley
ENGLISH LITERATURE Jonathan Bate
ENVIRONMENTAL ECONOMICS Stephen Smith
EPIDEMIOLOGY Rodolfo Saracci
ETHICS Simon Blackburn
THE EUROPEAN UNION John Pinder and Simon Usherwood
EVOLUTION Brian and Deborah Charlesworth
EXISTENTIALISM Thomas Flynn
FASCISM Kevin Passmore
FASHION Rebecca Arnold
FEMINISM Margaret Walters
FILM Michael Wood
FILM MUSIC Kathryn Kalinak
THE FIRST WORLD WAR Michael Howard
FOLK MUSIC Mark Slobin
FORENSIC PSYCHOLOGY David Canter
FORENSIC SCIENCE Jim Fraser
FOSSILS Keith Thomson

FOUCAULT Gary Gutting
FREE SPEECH Nigel Warburton
FREE WILL Thomas Pink
FRENCH LITERATURE John D. Lyons
THE FRENCH REVOLUTION William Doyle
FREUD Anthony Storr
FUNDAMENTALISM Malise Ruthven
GALAXIES John Gribbin
GALILEO Stillman Drake
GAME THEORY Ken Binmore
GANDHI Bhikhu Parekh
GENIUS Andrew Robinson
GEOGRAPHY John Matthews and David Herbert
GEOPOLITICS Klaus Dodds
GERMAN LITERATURE Nicholas Boyle
GERMAN PHILOSOPHY Andrew Bowie
GLOBAL CATASTROPHES Bill McGuire
GLOBAL ECONOMIC HISTORY Robert C. Allen
GLOBAL WARMING Mark Maslin
GLOBALIZATION Manfred Steger
THE GOTHIC Nick Groom
GOVERNANCE Mark Bevir
THE GREAT DEPRESSION AND THE NEW DEAL Eric Rauchway
HABERMAS James Gordon Finlayson
HEGEL Peter Singer
HEIDEGGER Michael Inwood
HERODOTUS Jennifer T. Roberts
HIEROGLYPHS Penelope Wilson
HINDUISM Kim Knott
HISTORY John H. Arnold
THE HISTORY OF ASTRONOMY Michael Hoskin
THE HISTORY OF LIFE Michael Benton
THE HISTORY OF MATHEMATICS Jacqueline Stedall
THE HISTORY OF MEDICINE William Bynum
THE HISTORY OF TIME Leofranc Holford-Strevens
HIV/AIDS Alan Whiteside
HOBBES Richard Tuck
HUMAN EVOLUTION Bernard Wood
HUMAN RIGHTS Andrew Clapham
HUMANISM Stephen Law
HUME A. J. Ayer
IDEOLOGY Michael Freeden
INDIAN PHILOSOPHY Sue Hamilton
INFORMATION Luciano Floridi
INNOVATION Mark Dodgson and David Gann
INTELLIGENCE Ian J. Deary
INTERNATIONAL MIGRATION Khalid Koser
INTERNATIONAL RELATIONS Paul Wilkinson
ISLAM Malise Ruthven
ISLAMIC HISTORY Adam Silverstein
ITALIAN LITERATURE Peter Hainsworth and David Robey
JESUS Richard Bauckham
JOURNALISM Ian Hargreaves
JUDAISM Norman Solomon
JUNG Anthony Stevens
KABBALAH Joseph Dan
KAFKA Ritchie Robertson
KANT Roger Scruton
KEYNES Robert Skidelsky
KIERKEGAARD Patrick Gardiner
THE KORAN Michael Cook
LANDSCAPES AND GEOMORPHOLOGY Andrew Goudie and Heather Viles
LANGUAGES Stephen R. Anderson
LATE ANTIQUITY Gillian Clark
LAW Raymond Wacks
THE LAWS OF THERMODYNAMICS Peter Atkins
LEADERSHIP Keith Grint
LINCOLN Allen C. Guelzo
LINGUISTICS Peter Matthews
LITERARY THEORY Jonathan Culler
LOCKE John Dunn
LOGIC Graham Priest
MACHIAVELLI Quentin Skinner
MADNESS Andrew Scull
MAGIC Owen Davies
MAGNA CARTA Nicholas Vincent
MAGNETISM Stephen Blundell
MALTHUS Donald Winch
MAO Delia Davin
MARINE BIOLOGY Philip V. Mladenov
THE MARQUIS DE SADE John Phillips
MARTIN LUTHER Scott H. Hendrix
MARTYRDOM Jolyon Mitchell
MARX Peter Singer
MATHEMATICS Timothy Gowers

THE MEANING OF LIFE Terry Eagleton
MEDICAL ETHICS Tony Hope
MEDICAL LAW Charles Foster
MEDIEVAL BRITAIN John Gillingham and Ralph A. Griffiths
MEMORY Jonathan K. Foster
METAPHYSICS Stephen Mumford
MICHAEL FARADAY Frank A. J. L. James
MODERN ART David Cottington
MODERN CHINA Rana Mitter
MODERN FRANCE Vanessa R. Schwartz
MODERN IRELAND Senia Pašeta
MODERN JAPAN Christopher Goto-Jones
MODERN LATIN AMERICAN LITERATURE Roberto González Echevarría
MODERNISM Christopher Butler
MOLECULES Philip Ball
THE MONGOLS Morris Rossabi
MORMONISM Richard Lyman Bushman
MUHAMMAD Jonathan A. C. Brown
MULTICULTURALISM Ali Rattansi
MUSIC Nicholas Cook
MYTH Robert A. Segal
THE NAPOLEONIC WARS Mike Rapport
NATIONALISM Steven Grosby
NELSON MANDELA Elleke Boehmer
NEOLIBERALISM Manfred Steger and Ravi Roy
NETWORKS Guido Caldarelli and Michele Catanzaro
THE NEW TESTAMENT Luke Timothy Johnson
THE NEW TESTAMENT AS LITERATURE Kyle Keefer
NEWTON Robert Iliffe
NIETZSCHE Michael Tanner
NINETEENTH-CENTURY BRITAIN Christopher Harvie and H. C. G. Matthew
THE NORMAN CONQUEST George Garnett
NORTH AMERICAN INDIANS Theda Perdue and Michael D. Green
NORTHERN IRELAND Marc Mulholland
NOTHING Frank Close
NUCLEAR POWER Maxwell Irvine
NUCLEAR WEAPONS Joseph M. Siracusa
NUMBERS Peter M. Higgins
OBJECTIVITY Stephen Gaukroger
THE OLD TESTAMENT Michael D. Coogan
THE ORCHESTRA D. Kern Holoman
ORGANIZATIONS Mary Jo Hatch
PAGANISM Owen Davies
PARTICLE PHYSICS Frank Close
PAUL E. P. Sanders
PENTECOSTALISM William K. Kay
THE PERIODIC TABLE Eric R. Scerri
PHILOSOPHY Edward Craig
PHILOSOPHY OF LAW Raymond Wacks
PHILOSOPHY OF SCIENCE Samir Okasha
PHOTOGRAPHY Steve Edwards
PLAGUE Paul Slack
PLANETS David A. Rothery
PLANTS Timothy Walker
PLATO Julia Annas
POLITICAL PHILOSOPHY David Miller
POLITICS Kenneth Minogue
POSTCOLONIALISM Robert Young
POSTMODERNISM Christopher Butler
POSTSTRUCTURALISM Catherine Belsey
PREHISTORY Chris Gosden
PRESOCRATIC PHILOSOPHY Catherine Osborne
PRIVACY Raymond Wacks
PROBABILITY John Haigh
PROGRESSIVISM Walter Nugent
PROTESTANTISM Mark A. Noll
PSYCHIATRY Tom Burns
PSYCHOLOGY Gillian Butler and Freda McManus
PURITANISM Francis J. Bremer
THE QUAKERS Pink Dandelion
QUANTUM THEORY John Polkinghorne
RACISM Ali Rattansi
RADIOACTIVITY Claudio Tuniz
RASTAFARI Ennis B. Edmonds
THE REAGAN REVOLUTION Gil Troy
REALITY Jan Westerhoff

THE REFORMATION Peter Marshall
RELATIVITY Russell Stannard
RELIGION IN AMERICA Timothy Beal
THE RENAISSANCE Jerry Brotton
RENAISSANCE ART Geraldine A. Johnson
RHETORIC Richard Toye
RISK Baruch Fischhoff and John Kadvany
RIVERS Nick Middleton
ROBOTICS Alan Winfield
ROMAN BRITAIN Peter Salway
THE ROMAN EMPIRE Christopher Kelly
THE ROMAN REPUBLIC David M. Gwynn
ROMANTICISM Michael Ferber
ROUSSEAU Robert Wokler
RUSSELL A. C. Grayling
RUSSIAN HISTORY Geoffrey Hosking
RUSSIAN LITERATURE Catriona Kelly
THE RUSSIAN REVOLUTION S. A. Smith
SCHIZOPHRENIA Chris Frith and Eve Johnstone
SCHOPENHAUER Christopher Janaway
SCIENCE AND RELIGION Thomas Dixon
SCIENCE FICTION David Seed
THE SCIENTIFIC REVOLUTION Lawrence M. Principe
SCOTLAND Rab Houston
SEXUALITY Véronique Mottier
SHAKESPEARE Germaine Greer
SIKHISM Eleanor Nesbitt
THE SILK ROAD James A. Millward
SLEEP Steven W. Lockley and Russell G. Foster
SOCIAL AND CULTURAL ANTHROPOLOGY John Monaghan and Peter Just
SOCIALISM Michael Newman
SOCIOLOGY Steve Bruce
SOCRATES C. C. W. Taylor
THE SOVIET UNION Stephen Lovell
THE SPANISH CIVIL WAR Helen Graham
SPANISH LITERATURE Jo Labanyi
SPINOZA Roger Scruton
SPIRITUALITY Philip Sheldrake
STARS Andrew King
STATISTICS David J. Hand
STEM CELLS Jonathan Slack
STUART BRITAIN John Morrill
SUPERCONDUCTIVITY Stephen Blundell
SYMMETRY Ian Stewart
TERRORISM Charles Townshend
THEOLOGY David F. Ford
THOMAS AQUINAS Fergus Kerr
THOUGHT Tim Bayne
TOCQUEVILLE Harvey C. Mansfield
TRAGEDY Adrian Poole
THE TROJAN WAR Eric H. Cline
TRUST Katherine Hawley
THE TUDORS John Guy
TWENTIETH-CENTURY BRITAIN Kenneth O. Morgan
THE UNITED NATIONS Jussi M. Hanhimäki
THE U.S. CONGRESS Donald A. Ritchie
THE U.S. SUPREME COURT Linda Greenhouse
UTOPIANISM Lyman Tower Sargent
THE VIKINGS Julian Richards
VIRUSES Dorothy H. Crawford
WITCHCRAFT Malcolm Gaskill
WITTGENSTEIN A. C. Grayling
WORK Stephen Fineman
WORLD MUSIC Philip Bohlman
THE WORLD TRADE ORGANIZATION Amrita Narlikar
WRITING AND SCRIPT Andrew Robinson

Available soon:

HAPPINESS Daniel M. Haybron
THE PALESTINIAN-ISRAELI CONFLICT Martin Bunton
MANAGEMENT John Hendry
FOOD John Krebs
MODERN WAR Richard English

Mark Maslin

CLIMATE

A Very Short Introduction

UNIVERSITY PRESS

Great Clarendon Street, Oxford, OX2 6DP,
United Kingdom

Oxford University Press is a department of the University of Oxford.
It furthers the University's objective of excellence in research, scholarship,
and education by publishing worldwide. Oxford is a registered trade mark of
Oxford University Press in the UK and in certain other countries

First Edition published in 2013
Impression: 1

British Library Cataloguing in Publication Data
Data available

ISBN 978-0-19-964113-0

Printed in Great Britain by
Ashford Colour Press Ltd, Gosport, Hampshire

Contents

Acknowledgements

The author would like to thank the following people: Anne, Chris, Johanna, Alexandra, Abbie Maslin, and Sue Andrews for being there; Emma Marchant and Latha Menon for their excellent editing and support; all the staff and friends at the UCL Environment Institute, UCL Department of Geography, TippingPoint, Rezatec Ltd, Permian, DMCii, KMatrix, and Global Precious Commodities; and last but not least Miles Irving for excellent illustrations.

List of illustrations

Chapter 1
What is climate?

Introduction

Climate affects everything we do in life, from the clothes we wear to the diseases we catch. This is because as humans we only feel comfortable within a very narrow range of temperature and humidity. This comfort zone ranges from about 20°C to 26°C and from 20 to 75 per cent relative humidity (see Figure 1). However, we live almost everywhere in the world, meaning that conditions are frequently outside this comfort zone, and we have learnt to adapt our clothing and dwellings to maintain our comfort. So while you may think the clothes you have hanging in your wardrobe simply reflect your fashion taste or lack of, in reality they reflect the climate in which you live and how it changes throughout the year. So you have a thick padded coat for a Canadian winter and a short-sleeved shirt for a business meeting in Rio. Our wardrobes also give hints about where we like to take our holidays. If you are a budding Polar explorer then there will very warm Arctic clothes hanging up—if you love sunning yourself on the beach, then there will be shorts or a bikini instead.

Our houses are also built with a clear understanding of local climate. In England almost all houses have central heating as the outside temperature is usually below 20°C, but few have air conditioning as temperatures rarely exceed 26°C. On the other

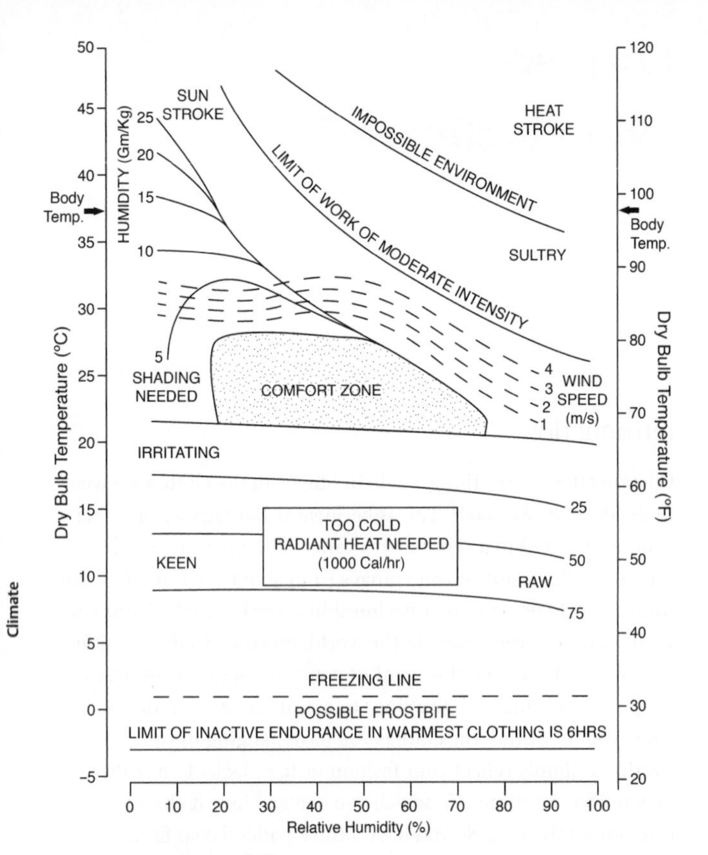

1. Human comfort and climate

hand, in Australia most houses have air conditioning but rarely central heating. Climate also affects the structure of our cities and how transport systems around the world operate. In Houston, Texas, there is a network of 7 miles of underground tunnels connecting all the major downtown buildings; this is fully climate controlled and links 95 heavily populated city blocks. People use the tunnel when it is raining or hot outside, because for at least 5 months of the year the average temperature in Houston is above

30 °C. Similarly there are underground malls in Canada to avoid the problems of heavy snow and extreme cold.

Climate controls where and when we get our food, because agriculture is controlled by rainfall, frost, and snow, and by how long the growing season is, which includes both the amount of sunlight and the length of the warm season. So in a simplified way, rice is grown where it is warm and very wet, while wheat can grow in much more temperate climes. The climate can also affect the quality of our food, for example it is well known that the very best vintages of French wine are produced when there are a few short sharp frosts during the winter, which harden the vines, producing excellent grapes. Farmers can also 'help' the local climate, for example by growing tomatoes in a greenhouse or by irrigating the land to provide a more constant supply of water.

Climate also influences where there will be extreme weather events such as heatwaves, droughts, floods, and storms. However in many cases our perception of extreme events is determined by local conditions, so for example in 2003 northern Europe was hit with a 'heatwave' and 100 °F (37.8 °C) was recorded for the first time ever in England. However in countries of the tropics a heatwave would not be recorded until temperatures were above 45 °C. Climate also has a large effect on our health, as many diseases are temperature and humidity controlled. For example incidences of influenza, commonly called the flu, reach a peak in winter. Since the Northern and Southern Hemispheres have winter at different times of the year, there are actually two different flu seasons globally each year. The influenza virus migrates between the two hemispheres after each winter, giving us time to produce new vaccinations based on the new strain of flu that has appeared in the previous six months in the other hemisphere. There have been many arguments about why flu is climate controlled and the theory is that during cold dry conditions the virus can survive on surfaces longer and so be more easily transmitted between people. Another suggestion is that vitamin D might provide some resistance or immunity to the virus. Hence in winter and during the tropical rainy season, when people stay indoors, away from the sun, their vitamin D levels fall and incidences of influenza increase.

Hot and cold Earth

The climate of our planet is caused by the Equator of the Earth receiving more of the sun's energy than the poles. If you imagine the Earth is a giant ball, the closest point to the sun is the middle or the Equator. The Equator is where the sun is most often directly overhead and it is here that the Earth receives the most energy. As you move further north or south away from the Equator, the surface of the Earth curves away from the sun, increasing the angle of the surface of the Earth relative to the sun. This means the sun's energy is spread over a larger area, and thus causes less warming. If we lived on a flat disc we would get much more energy from the

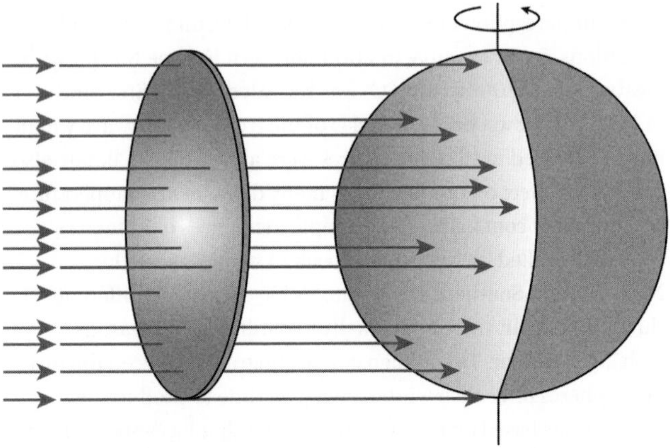

2. Solar energy distributed over a sphere

sun—about 1,370 Watts per square metre (W/m^2)—instead the planet surface averages about 343 W/m^2 (Figure 2) due to its curved nature. The Earth also receives a very small fraction of the energy pumped out of the sun. If you consider how small the Earth is compared with the sun, for every Watt we receive from the sun, it emits 2 billion Watts. This is why in many science fiction novels the authors imagine a strip or even a sphere around a star to collect all that energy that is simply being lost into space.

About one-third of the solar energy we receive is reflected straight back into space. This is because of 'albedo', which means how reflective is a surface. So, for example, white clouds and snow have a very high albedo and reflect almost all of the sunlight that falls on them, while darker surfaces such as the oceans, grassland, and rainforest absorb a lot more energy. Not only do the poles receive less energy than the Equator, but they also lose more energy back into space (Figure 3): the white snow and ice in the Arctic and Antarctic

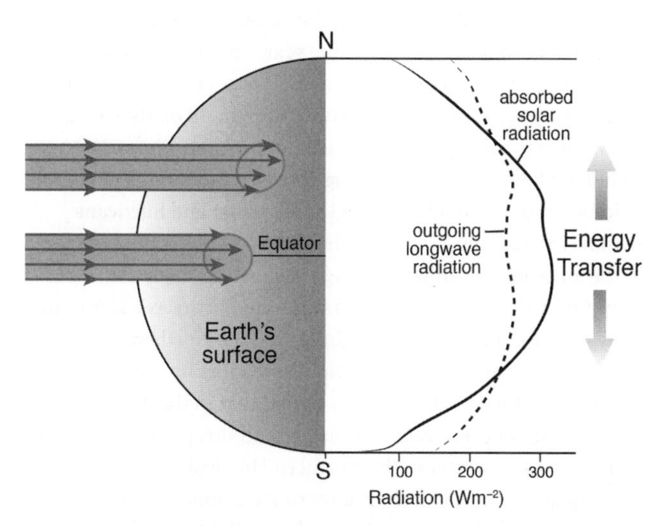

3. Energy transfer away from the Equator driven by angle of solar radiation

have a high albedo and bounce a lot of the sun energy back into space. On the other hand, the darker much less reflective vegetation at lower latitudes absorb a lot more energy. These two processes working together mean that the tropics are hot and the poles are very cold. Nature hates this sort of energy imbalance, so energy, in the form of heat, is transported by both the atmosphere and oceans from the Equator to both poles, and this affects the climate.

Earth in space

Our climate is controlled by two fundamental facts about the relationship between the Earth and the sun. The first is the tilt of the Earth's axis of rotation, which causes the seasons. The second factor is the daily rotation of the Earth that provides us with night and day and drives the circulation of both the atmosphere and the oceans.

The Earth's axis of rotation is tilted at an angle of 23.5° and results in a seasonal difference in the amount of energy received by each hemisphere throughout the year. The seasonal changes are by far the largest effect on climate. It is amazing to think that if the Earth were not tilted and stood straight up on its axis then we would not have spring, summer, autumn, and winter. We would not have the massive change in vegetation in the temperate latitudes and we would not have the monsoon and hurricane seasons in the tropics. The reason for the seasons is the change in the angle of the sunlight hitting the Earth through the year. If we take 21 December as an example, the Earth's axis is leaning away from the sun, so the sunlight hitting the Northern Hemisphere is at a greater angle and spreading its energy over a wider area. Moreover the lean is so great that in the Arctic the sunlight cannot even reach the surface and this produces 24 hours of darkness and winter in the Northern Hemisphere. However, everything is opposite in the Southern Hemisphere, since it is then leaning towards the sun and hence the sunlight is more directly overhead. This means that Antarctica is bathed in 24 hours of sunlight and people in Australia have Christmas dinner on the beach, while topping up their tan. As the Earth moves round the sun, taking about 365.25 days (hence the leap year every fourth year), the angle of the axis stays in the same place. Hence when it comes to June the Earth's axis is leaning towards the sun, so the Northern Hemisphere has lots of direct sunlight and thus summer, while the Southern Hemisphere is shielded from the sunlight and is plunged into winter.

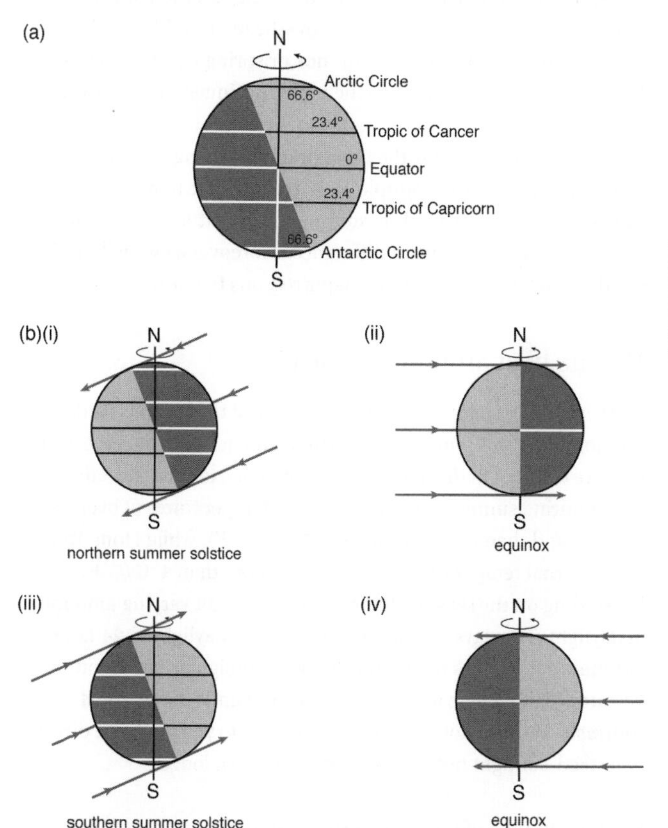

4. Solstice and equinox caused by the tilt of the Earth

If we follow the sun through a year we can see how this tilt affects the Earth through the seasons. If we start at 21 June the sun is overhead at midday at the Tropic of Cancer (23.4°N), the northern summer solstice. The angle of the sun moves southward until 21 September when it is overhead at midday over the Equator, the equinox or autumn equinox in the Northern Hemisphere. The sun appears to continue southward and on 21 December it is overhead at midday at the Tropic of Capricorn (23.4°S) the southern summer solstice. The sun then appears to move northward until it is directly overhead at midday at the Equator on the 21 March the equinox or spring equinox in the Northern Hemisphere and so the cycle continues (see Figure 4).

The seasons signal by far the most dramatic change in our climate; if we take for example New York, winter temperatures can be as low as −20°C while summer temperatures can be over 35°C—a 55°C temperature difference. Moreover as we will find out the seasons are one of the major reasons for storms.

Moving heat around the Earth

The second big factor affecting the climate of the Earth is its daily rotation. First this plunges the Earth in and out of darkness causing massive changes in diurnal temperature. For example the Sahara desert during summer can have daytime temperatures of over 38°C (100°F) and then nighttime lows of 5°C (40°F); while Hong Kong has a diurnal temperature range of little more than 4°C (7°F). Depending on the season, different areas also get varying amounts of daylight. The days can vary from 24 hours' daylight to 24 hours' darkness at the poles to around 12 hours' sunlight every day at the Equator. This change in the daylight compounds the seasonal contrasts, because not only during summer do you get more direct 'overhead' sunlight but also you get it for much longer.

But the spinning of the Earth also makes the transport of heat away from the Equator more complicated. This is because the spinning of the Earth makes everything else including the atmosphere and oceans turn. The simple rule is that rotation of the Earth causes the air and ocean currents to be pushed to the right of the direction they are travelling in the Northern Hemisphere and to the left of the direction they are travelling in the Southern Hemisphere. This deflection is called the Coriolis effect and its strength increases the further you go towards the poles.

An everyday example of this, which is always quoted, is the way water flows down a plughole or a toilet. In the Northern Hemisphere water is said to flow clockwise down the plughole while in the Southern Hemisphere it is anti-clockwise. However, I hate to tell you that the direction the water drains out of your bath or toilet is not related to the Coriolis effect or to the rotation of the Earth. Moreover no consistent difference in rotation direction between toilets in the Northern and Southern Hemispheres has been observed. This is because the Coriolis effect has such a small influence compared with any residual movement of the water and the effect of the shape of the container. This also mean the wonderful cottage industry of communities living on the Equator showing tourists the Coriolis effect is simply done by a sleight of hand. For example in Kenya there are big signs up telling you when you are crossing the Equator; if you care to stop at the road side locals will happily pour water from a bucket into a large funnel and seeming to demonstrate clearly that it goes a different way round when you are standing on one side of the sign than when you are standing on the other. However, this change is all in the wrist and how the water is poured in, affecting which way it goes round. Still, even though it is completely fake, I love these demonstrations as it means loads of locals and tourists get to hear about the Coriolis effect!

Back to climate, so why do the ocean currents and winds have this deflection? Imagine firing a missile from the Equator directly north. Because the missile was fired from the Earth which is spinning eastward, the missile is also moving east. As the Earth spins the Equator has to move fast through space to keep up with the rest, as it is the widest part of the Earth. As you go further north or south away from the Equator the surface of the Earth curves in, so it does not have to move as fast to keep up with the Equator. So in one day the Equator must move round 40,074 km (the diameter of the Earth) a speed of 1,670 km/hour, while the Tropic of Cancer (23.4°N) has to move 36,750 km, with a speed of 1,530 km/hour, and the Arctic circle (66.6°N) has to move 17,662 km so has a speed of 736 km/hour. At the North Pole there is no relative movement at all so the speed there is 0 km/hour. A practical demonstration of this is if you hold hands with a friend and stand in the same place while spining them around, they will travel much faster than you do. Therefore the missile, fired from the Equator, has the faster eastward speed of the Equator; as it moves northward towards the Tropic of Cancer, the surface of the Earth is not moving as fast eastward as the missile. This gives the appearance that the missile is moving northeast as it is moving faster eastward than the area it is moving into. Of course the closer you get to the poles the greater this difference in speeds so the greater the deflection to the east (see Figure 5).

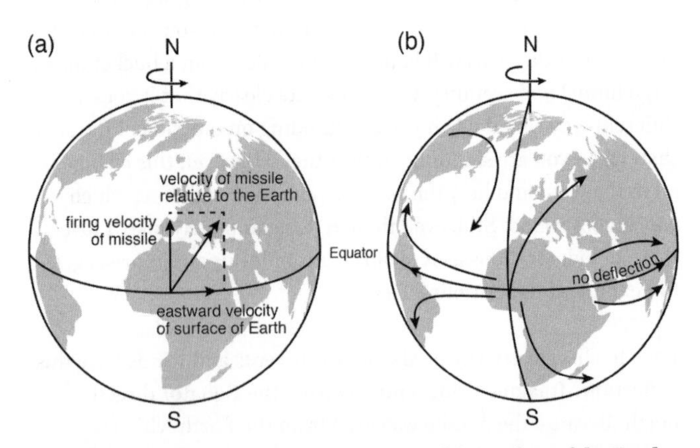

5. Coriolis effect due to relative movement of the surface of the Earth

Summary

The climate system is very straightforward. It is controlled by the different amount of solar energy received at the Equator and the poles. Climate is simply the redistribution of energy to undo this imbalance. It is the atmosphere and the oceans which undertake this redistribution, as we will see in Chapter 2. Complications are added because the Earth's axis of rotation is at an angle with respect to the sun, which leads to there being a strong season cycle. On top of this the Earth rotates every 24 hours, plunging the Earth in and out of darkness. It also means the redistribution of energy away from the Equator takes place on a spinning ball. This creates the Coriolis effect and helps to explain why nearly all weather systems seem to spin.

Chapter 2
Atmosphere and oceans

This chapter examines the effects of both the atmosphere and the oceans on climate and how they store and redistribute solar heat around the globe. It will explain why the ocean dominates in the movement of heat away from the Equator while the atmosphere dominates in the mid- to high latitudes. The chapter will finish by summarizing the major climate zones of the world and explaining why there are globally three main rain belts and two main desert belts.

The atmosphere

The atmosphere is the home of our weather. It begins at the surface of the Earth and becomes thinner and thinner with increasing altitude, with no definite boundary between the atmosphere and outer space. The arbitrary Kármán line at 62 miles (100 km), named after Theodore von Kármán (1881–1963), a Hungarian-American engineer and physicist, is usually used to mark the boundary between atmosphere and outer space. The layer of atmosphere in which weather takes place is thinner at about 10 miles thick. The oceans also play an important part in controlling our weather and climate. The oceans are on average about 2.5 miles deep, so the total thickness of the layer controlling our climate is 12.5 miles thick.

The atmosphere is a mechanical mixture of gases, not a chemical compound. What is significant is that these gases are mixed in remarkably constant proportions up to about 50 miles (80 km) above the surface of the Earth. Four gases, nitrogen, oxygen, argon, and carbon dioxide account for 99.98 per cent of air by volume. Of special interest are the greenhouse gases that despite their relative scarcity have a great effect on the thermal properties of the atmosphere, which include carbon dioxide, methane, and water vapour. The greenhouse effect is discussed later in this chapter while global warming is discussed at greater length in Chapter 8.

Content of the atmosphere

Nitrogen is a colourless, odourless, tasteless, and mostly inert gas and makes up ~78 per cent by volume of the Earth's atmosphere. Argon is also a colourless, odourless, tasteless, and completely inert gas and makes up ~0.9 per cent by volume of the Earth's atmosphere. In contrast oxygen is a very reactive gas and makes up ~21 per cent of the Earth's atmosphere by volume. Oxygen sustains all life on Earth and is constantly recycled between the atmosphere and the biological processes of plants and animals. It combines with hydrogen to produce water, which in its gaseous state, water vapour, is one of the most important components of the atmosphere as far as weather is concerned.

Oxygen also forms another gas called ozone or trioxygen, which is made up of three oxygen atoms instead of the usual two. This is an extremely important gas in the atmosphere as it forms a thin layer in the stratosphere (between 6 and 31 miles) that filters out harmful ultraviolet radiation that can cause cancer. However, even in this 'layer' the ozone concentrations are only two to eight parts per million in volume, so most of the oxygen remains of the normal dioxygen type. Much of this important gas was being destroyed by our use of CFCs, and ozone holes have been found over the Arctic and Antarctic, until governments worldwide agreed (for example, in the Vienna Convention for the Protection of the Ozone Layer in 1985 and then in the Montreal Protocol on Substances that Deplete the Ozone Layer in 1987) to stop the use of all CFCs and related compounds.

Carbon dioxide makes up 0.04 per cent of the Earth's atmosphere and is a major greenhouse gas, important for keeping the Earth relatively warm. Until recently, the level of carbon dioxide has been balanced through its consumption by plants for photosynthesis and its production by plants and animals in respiration. However, human industry over the last 100 years has caused a lot more carbon dioxide to be pumped out into the atmosphere, upsetting this natural balance.

Aerosols are suspended particles of sea salt, dust (particularly from desert regions), organic matter, and smoke. The height at which these aerosols are introduced will determine whether they cause regional warming or regional cooling. This is because high up in the atmosphere they help reflect sunlight thus cooling the local area, while at low altitudes they absorb some of the warmth coming off the Earth thus warming the local air. Industrial processes have increased the level of aerosols in the atmosphere, which has lead to smog in urban areas, acid rain, and localized cooling causing 'global dimming'. But the most important effect of aerosols is to help clouds form. Without these minute particles water vapour cannot condense and form clouds; and without cloud precipitation there is no weather.

Water vapour is the forgotten but most important greenhouse gas, which makes up about 1 per cent by volume of the atmosphere, but is highly variable in time and space as it is tied to the complex global hydrological cycle. The most important role that water vapour plays in the atmosphere is the formation of clouds and the production of precipitation (rain or snow). Warm air can hold more water vapour than cold air. So whenever a parcel of air is cooled down, for example as air rises or meets a cold air mass, it cannot hold as much water vapour, so the water condenses on to aerosols and produces clouds. An important point which we discuss later is that as water changes from a gas to a liquid it releases some energy, and it is this energy which can fuel storms as large as hurricanes. Clouds come in all sorts of shapes and sizes and are an excellent way of telling what sort of weather is coming up!

Greenhouse effect

The temperature of the Earth is determined by the balance between energy from the sun and its loss back into space. Of Earth's incoming solar short-wave radiation (mainly ultraviolet radiation and visible 'light'), nearly all of it passes through the atmosphere without interference (Figure 6). The only exception is ozone, which luckily for us absorbs energy in the high-energy UV band (which is very damaging to our cells), restricting how much reaches the surface of the Earth. About one-third of the solar energy is reflected straight back into space. The remaining energy is absorbed by both the land and ocean, which warms them up. They then radiate this acquired warmth as long-wave infrared or 'heat' radiation. Atmospheric gases such as water vapour, carbon dioxide, methane, and nitrous oxide are known as greenhouse gases as they can absorb some of this long-wave radiation, thus warming the atmosphere. This effect has been measured in the atmosphere and can be reproduced time and time again in the laboratory. We need this greenhouse effect because without it, the Earth would be at least 35 °C colder, making the average temperature in the tropics about –5 °C. Since the Industrial Revolution we have been burning fossil fuels (oil, coal, natural gas) deposited hundreds of millions years ago, releasing the carbon back into the atmosphere as carbon dioxide and methane, increasing the 'greenhouse effect', and elevating the temperature of the Earth. In effect we are releasing ancient stored sunlight back in to the climate system thus warming up the planet.

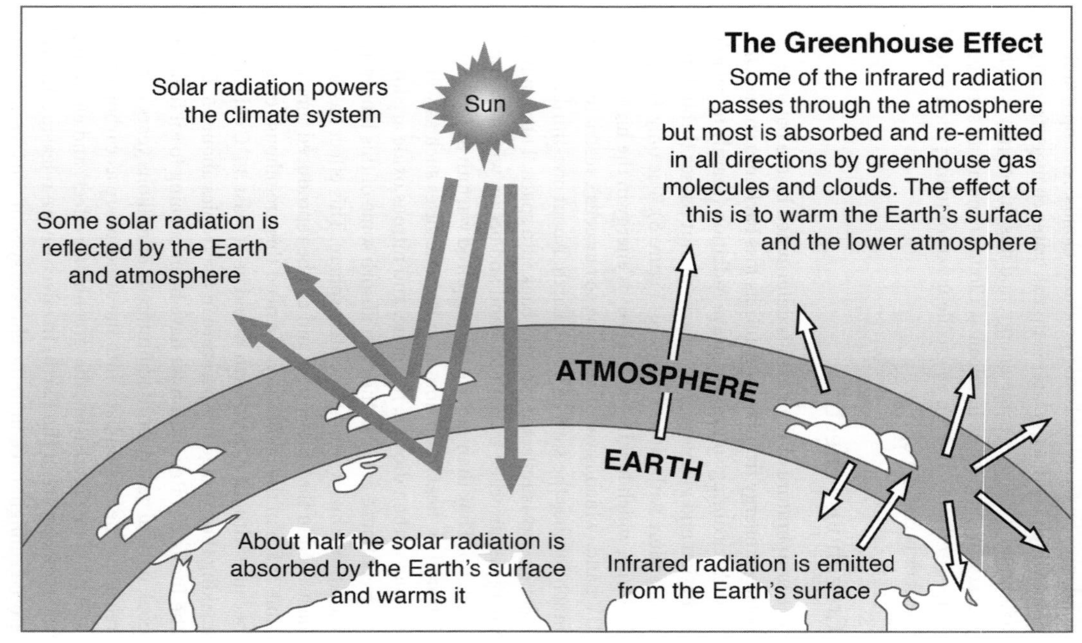
The Greenhouse Effect
Some of the infrared radiation passes through the atmosphere but most is absorbed and re-emitted in all directions by greenhouse gas molecules and clouds. The effect of this is to warm the Earth's surface and the lower atmosphere
Solar radiation powers the climate system
Sun
Some solar radiation is reflected by the Earth and atmosphere
ATMOSPHERE
EARTH
About half the solar radiation is absorbed by the Earth's surface and warms it
Infrared radiation is emitted from the Earth's surface

6. The greenhouse effect

Box 1 Vertical structure of the atmosphere

The atmosphere can be divided conveniently into a number of well demarcated horizons, mainly based on temperature (see Figure 7).

Troposphere

The lowest layer of the atmosphere is the zone where atmospheric turbulence and weather are most marked. It contains 75 per cent of the total molecular mass of the atmosphere and virtually all the water vapour. Throughout this layer there is a general decrease in temperature at a mean rate of 6.5°C/km, and the whole zone is

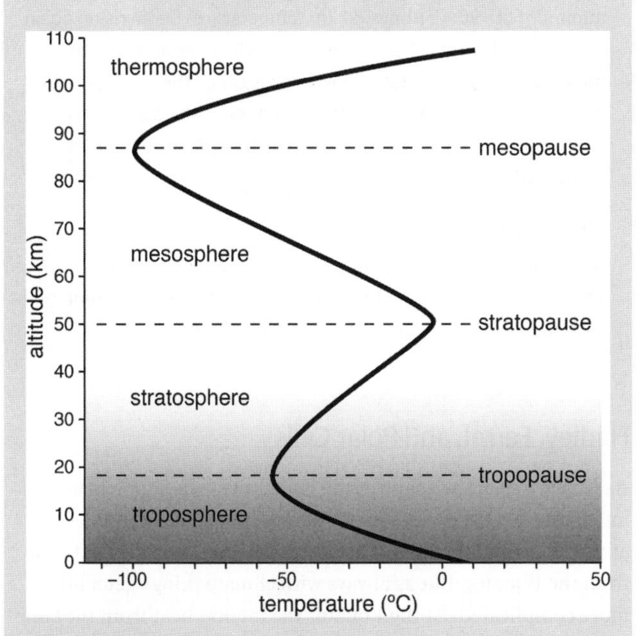

7. Temperature profile through the atmosphere

capped by a temperature inversion layer. This layer, called the 'tropopause', acts as a lid on the troposphere and on weather.

Stratosphere

The second major atmospheric layer extends upwards from the tropopause to about 50 km. Although the stratosphere contains much of the ozone, the maximum temperature caused by the absorption of ultraviolet radiation occurs at the 'stratopause' where temperatures may exceed 0 °C. This large temperature increase is due to the relative low density of the air at this height.

Mesosphere

Above the stratopause average temperatures decrease to a minimum of −90 °C. Above 80 km temperatures begin rising again because of absorption of radiation by both ozone and oxygen molecules. This temperature inversion is called the 'mesopause'. Pressure is extremely low in the mesosphere decreasing from 1 mb at 50 km to 0.01 mb at 90 km (surface pressure is about 1,000 mb).

Thermosphere

Above the mesopause, atmospheric densities are very low. Temperatures rise throughout this zone due to the absorption of solar radiation by molecular and atomic oxygen.

Hadley, Ferrel, and Polar Cells

As we have seen the shape of the Earth sets up a temperature imbalance between the Equator and the poles. Both the atmosphere and the oceans act as transporters for this heat away from the Equator. But as always with climate things get a little more complicated. At the Equator the intense heat from the sun warms up the air near the surface and causes it to rise high into the atmosphere. Warm air rises because the gas molecules in warm air can move further apart making the air less dense, and correspondingly cold air sinks. This loss of air upwards creates a space and low atmospheric pressure, which is filled by air being sucked in. This produces the Trade Winds in both the North and South Hemisphers. The northeast and southeast Trade Winds meet at the Inter-Tropical Convergence Zone (ITCZ). This causes a problem as the climate system is desperately trying to export heat away from the region around the Equator and these in-blowing winds do nothing to help this removal of heat. So in the tropics it is the surface currents of the ocean that transport most of the heat (see Figure 8). These currents include the Gulf Stream, which takes heat from the tropical Atlantic and transports it northward keeping Europe's weather mild all year round (see Box 2). Other currents include the Kuroshiro current in the western North Pacific, the Brazilian current in the western South Atlantic, and finally the East Australian current in the western South Pacific.

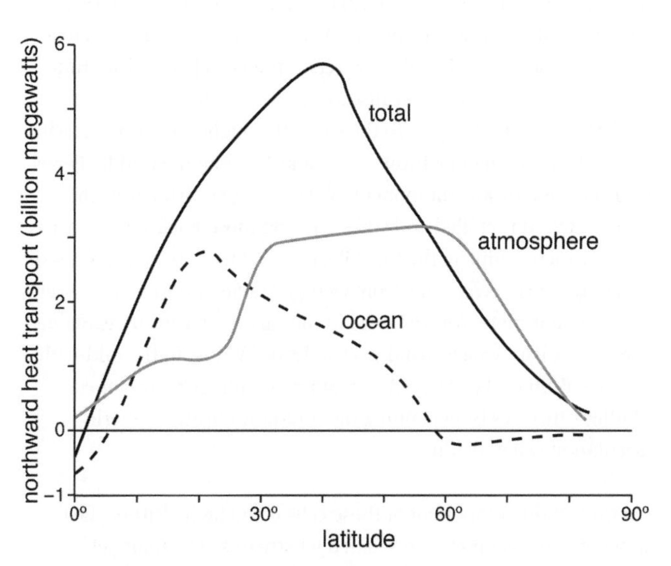

8. Heat transport away from the Equator

However, the hot air which has risen high into the atmosphere in the tropics slowly cools, due to both its rise and its movement towards the poles, and at about 30° north and south it sinks, forming the sub-tropical high pressure zone. As this sinking air reaches the surface it spreads out, moving both north and south. This sinking air has lost most of its moisture and therefore dries out the land it sinks onto, producing some of the major deserts around the world. The southward air links into the first atmospheric cell called the Hadley Cell and becomes part of the Trade Wind system. While the northward-bound air forms the Westerlies and it is from here northwards that the atmosphere takes over from the oceans as the major transporter of heat. The movement of warm sub-tropical air northward is only stopped when it meets the cold Polar air mass at the Polar Front. The intense cold at the poles causes air to become super chilled and sink, causing out-blowing winds (Figure 9). When this cold Polar air meets the warm, moist Westerlies at the Polar Front the clash causes the Westerlies to lose a lot of their moisture in the form of rain. It also forces the warm sub-tropical air to rise, as the cold Polar air is much heavier. This rising air completes the other two cells, the Ferrel or mid-latitude cell, and the Polar Cell—because as the air rises it spreads out to both the north and south. To the south this high-rise air meets with tropical air coming northward and sinks forming the middle Ferrel Cell. The northward component of this rising air drifts over the poles where it is chilled and sinks forming those Polar out-blowing winds which complete the third Polar Cell (Figure 9). The names of two of the three cells come from George Hadley, an English lawyer and amateur meteorologist, who in the early 18th century explained the mechanism which sustained the Trade Winds. In the mid-19th century William Ferrel, an American meteorologist, developed Hadley's theories by explaining the mid-latitude atmospheric circulation cell in detail.

An important component of these cells is the high altitude, fast flowing, narrow air currents called jet streams. The main jet streams are located near the tropopause, which represents the

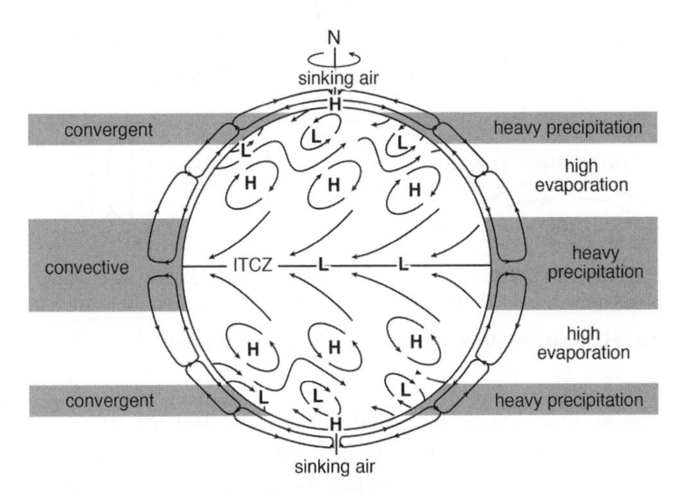

9. Major atmospheric circulation patterns

transition between the troposphere and the stratosphere (see Box 1). The major jet streams are westerly winds that flow west to east. Their paths typically have a meandering shape; jet streams may start, stop, split into two or more parts, combine into one stream, or flow in various directions including the opposite direction of most of the jet. The strongest jet streams are the polar jets, at around 7–12 km above sea level, and the higher and somewhat weaker sub-tropical jets at around 10–16 km (see Figure 10). The Northern and the Southern Hemispheres each have both a Polar jet and a sub-tropical jet. The Northern Hemisphere Polar jet flows over the middle to northern latitudes of North America, Europe, and Asia and their intervening oceans, while the Southern Hemisphere Polar jet mostly circles Antarctica all year round. Jet streams are caused by a combination of the Earth's rotation and energy in the atmosphere, hence they form near boundaries of air masses with significant differences in temperature (Figure 10).

Though the general wind patterns of Earth follow this simple three-celled, two jet stream per hemisphere model, in reality

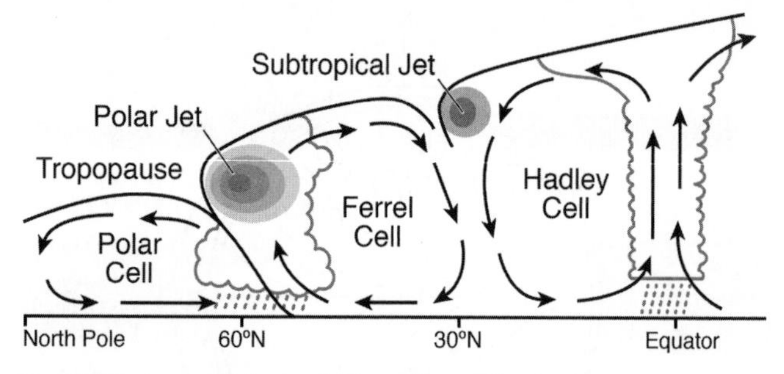

10. Major atmospheric circulation cells and jet streams

they are much more complicated. First because the Earth is spinning and this adds the influence of the Coriolis effect. This means that air masses trying to flow northward or southward are deflected by the spinning of the Earth. For example this causes large meanders in the jet streams, which are called planetary waves. These can have a huge effect on our weather, for instance in spring and summer 2012 the planetary waves within the Polar jet became fixed and brought a major heatwave to the USA and the wettest April, May, and June on record for England. Second, the continents heat up much quicker than the oceans, which can cause the surface air over the land to rise, which can alter the general circulation of surface wind. This can cause local land–sea breezes and, on a much larger scale, cause the monsoon systems. The seasons, then, can have a huge effect on atmospheric circulation, because during the summer in each hemisphere the land heats up much more than the ocean, hence the ITCZ is pulled southwards towards Australasia, and across South America and Southeast Africa during Southern Hemisphere summer and northwards across India, Southeast Asia and North Africa during Northern Hemisphere summer.

The Hadley Cells however do explain why there are three main rainfall belts across the Earth, the convection rainfall belt which moves north and south of the Equator and the two convergent rainfall belts one in the Northern and one in the Southern

Hemisphere where warm, moist sub-tropical air meets cold dry Polar air. They also explain why there are two main desert belts in the world, which are usually found between the rainfall belts with super dry air sinking between the Hadley and Ferrel Cells. In the Northern Hemisphere good examples are the Sahara desert in North Africa and the Gobi desert in China, while in the Southern Hemisphere, Central Australia and the Kalahari desert in South Africa are good examples.

The Hadley Cells can also be used to define the three main storm zones. First are 'winter storms' at the Polar Front. Second are the sub-tropical highs and the Trade Wind belt, which are the spawning ground for hurricanes. Third is the ITCZ, where the rapidly rising air cools and produces tropical thunderstorms with heavy rainfall, producing monsoons as it moves over the land (see Chapter 4 for more details).

Surface ocean circulation

As we have seen the surface ocean is important in transporting heat around the globe. The circulation of the oceans starts with the wind, because it is the action of the wind on the surface ocean that makes it move (Figure 11). As the wind blows on the surface water, the friction allows energy to be transferred from the winds to the surface water, leading to major currents. The wind energy is transferred to greater depths in the water column turbulence, which allows wind driven currents to be very deep. There are three main types of current flow: (a) Ekman motion or transport; (b) inertia currents; and (c) geostrophic currents.

Ekman motion or transport

Vagn Walfrid Ekman (3 May 1874–9 March 1954) was a Swedish oceanographer who calculated that with a constant wind over an ocean that was infinitely deep and infinitely wide with the same density, the Coriolis effect would be the only other force acting on the water column. The further away from the surface and the

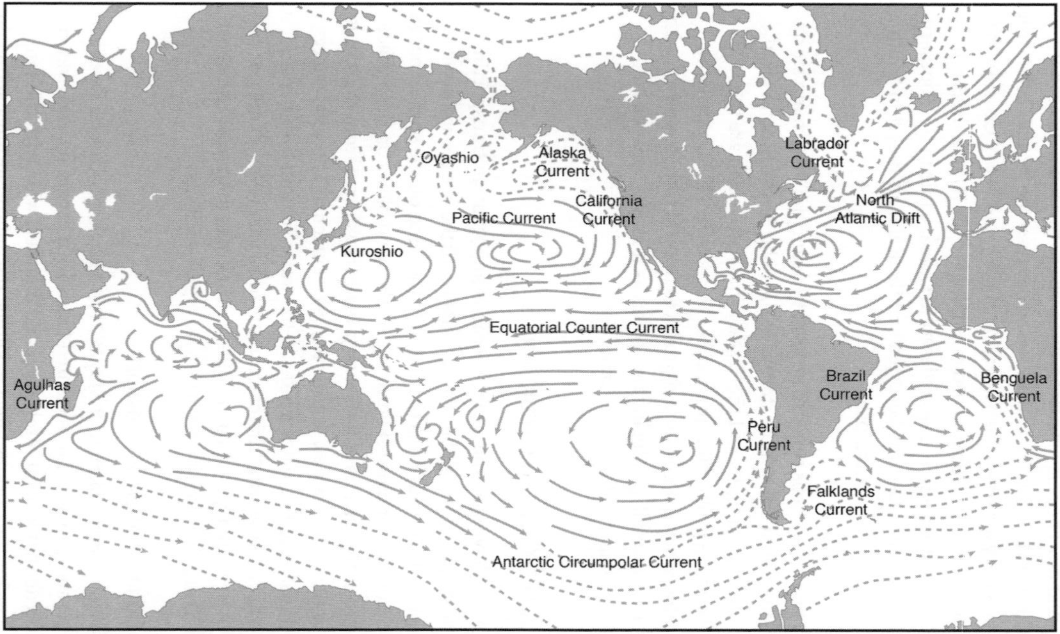

11. Major surface ocean current

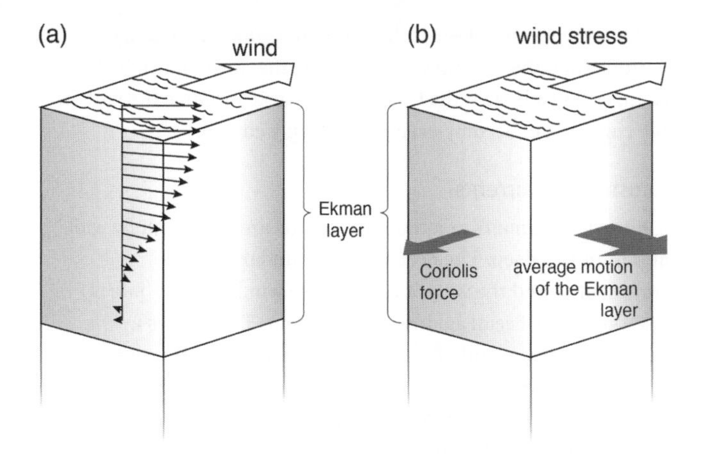

12. Ekman surface ocean movement due to wind action

diminishing influence of the wind, the greater the effect of Coriolis, which results in a spiral of water movement (see Figure 12). The result is that the net movement of the surface of the ocean is at 90 degrees to wind direction. This phenomenon was first noted by Fridtjof Nansen, during his arctic expeditions in the 1890s, when he recorded that ice transport appeared to occur at an angle to the wind direction. The direction of transport is of course dependent on the hemisphere. In the Northern Hemisphere this transport is at a 90° angle to the right of the direction of the wind, and in the Southern Hemisphere it occurs at a 90° angle to the left of the direction of the wind.

Inertia currents

Surface water masses are huge. For example, the Gulf Stream measures about 100 Sverdrup (Sv). One Sverdrup is $10^6 m^3/s$ or a million tonnes of water per second. The entire global input of freshwater from rivers to the ocean is equal to about 1 Sv. Hence these water masses have a huge momentum, and thus the currents continue to flow long after the wind has ceased pushing. When the wind stops blowing only friction and the Coriolis effect continues to act on the water mass. If the water mass does not change latitude then the current will flow along the line of latitude. If it changes latitude then the Coriolis effect acts and thus the path of the current will become even more steeply curved.

Geostrophic currents

Contrary to Ekman's assumptions, oceans are not infinitely wide and infinitely deep. The oceans have boundaries—the continents—and the water driven by the wind tends to 'pile up' on one side of the ocean against the continent. This causes a sea-surface slope, and affects the hydrostatic pressure with water flowing from areas of high to those of low pressure. This force is known as the horizontal pressure gradient force, and is also influenced by the Coriolis effect, producing what are known as geostrophic currents. One way of studying geostrophic currents is to look at the dynamic topography of the sea-surface—in other words, areas of the sea that are higher than the rest.

The combination of wind-blown Ekman currents, inertia currents, and geostrophic currents produces most of the major circulation features of the world's oceans (Figure 11). One of the major features is the gyres in each of the ocean basins. These large systems of rotating ocean currents are found in the North and South Atlantic Oceans, North and South Pacific Oceans, and the Indian Ocean. There is, however, another influence on surface ocean circulation and that is the pulling created by the sinking of surface water when deep-water currents are formed.

Deep-ocean circulation

The circulation of the deep ocean is one of the major controls on global climate due to its ability to exchange heat between the two hemispheres. In fact, the deep ocean is the only candidate for driving and sustaining internal long-term climate change (of hundreds to thousands of years) because of its volume, heat capacity, and inertia. Today the tropical sun heats the surface water in the Gulf of Mexico. This heat also causes there to be a lot of evaporation sending moisture into the atmosphere starting the hydrological cycle. All this evaporation leaves the surface water enriched in salt. So this hot salty surface water is pushed by the winds out of the Caribbean along the coast of Florida and into the North Atlantic Ocean. This is the start of the famous Gulf Stream (Figure 13). The Gulf Stream is about 500 times the size of the Amazon River at its widest point and flows along the coast of the USA and then across the North Atlantic Ocean, past the coast of Ireland, past Iceland, and up into the Nordic Seas. As the Gulf Stream flows northward it becomes the North Atlantic Drift and it cools down. The combination of a high salt content and low temperature increases the surface water density or heaviness.

Let us now examine the difference between freshwater and seawater. As freshwater is cooled down, something amazing

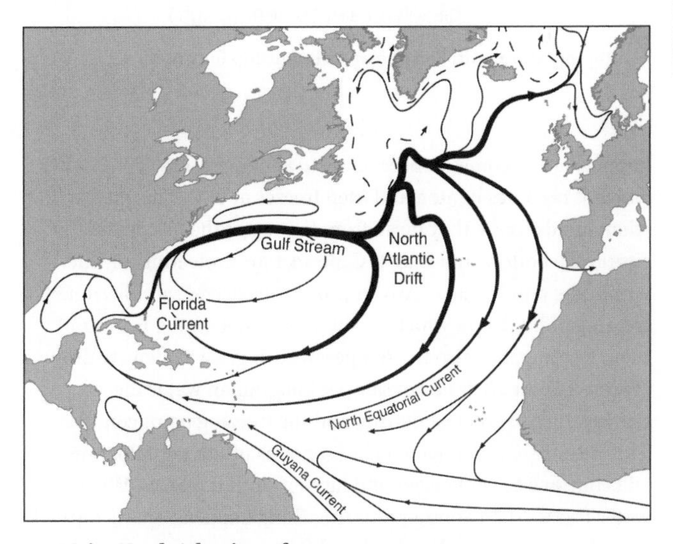

13. Major North Atlantic surface ocean currents

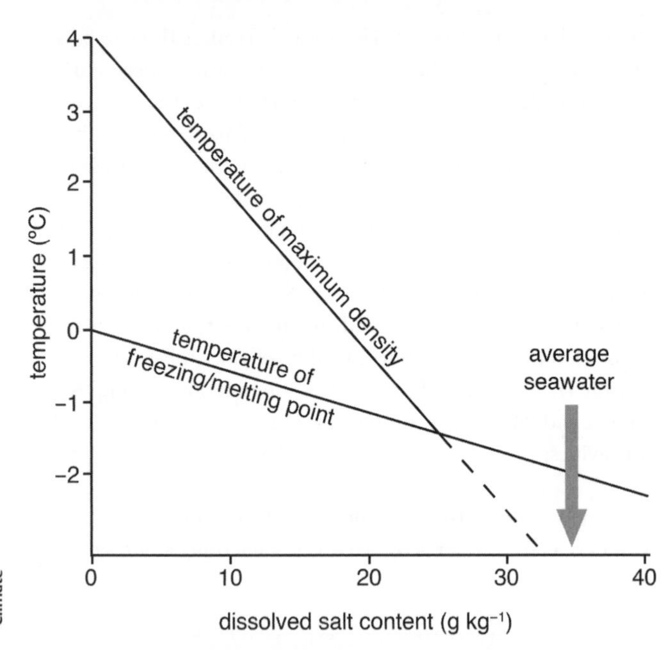

14. Temperature, salt, and density relationship for water

happens—it becomes denser down to a temperature of 4°C, after which it becomes lighter, and then freezes at 0°C. This means that when ponds freeze they do so from the top as the heaviest water sits on the bottom and is at 4°C, perfect for protecting any life within the pond or lake. As you progressively add salt to water, its freezing point drops, which is why we put salt on roads to stop them freezing, but also the temperature of greatest density drops (Figure 14). At 26 grams of salt per kilogram of water the temperature of greatest density and the freezing point coincide. This means seawater, which has 35 grams of salt per kilogram, will continue to get heavier and heavier until it freezes. When water freezes then another amazing thing happens—ice is formed, a solid that is lighter than its liquid form.

When the surface water reaches the relatively fresh oceans north of Iceland, the surface water has cooled sufficiently to become dense enough to sink into the deep ocean. The 'pull' exerted by the sinking of this dense water mass helps maintain the strength of the warm Gulf Stream, ensuring a current of warm tropical water flowing into the northeast Atlantic, sending mild air masses across to the European continent. It has been calculated that the Gulf Stream delivers the same amount of energy as a million nuclear power stations. If you are in any doubt about how good the Gulf Stream is for the European climate, compare the winters at the same latitude on either side of the Atlantic Ocean, for example London with Labrador, or Lisbon with New York. Or, better still, compare Western Europe and the west coast of North America, which have a similar geographical relationship between the ocean and continent—for example, Alaska and Scotland, which are at about the same latitude.

The newly formed deep water in the Nordic Seas sinks to a depth of between 2,000 metres and 3,500 metres in the ocean and flows southward down the Atlantic Ocean, as the North Atlantic Deep Water (NADW). In the South Atlantic Ocean, it meets a second type of deep water, which is formed in the Southern Ocean and is called the Antarctic Bottom Water (AABW). This is formed in a different way to NADW. Antarctica is surrounded by sea ice and deep water forms in coast polnyas (large holes in the sea ice). Out-blowing Antarctic winds push sea ice away from the continental edge to produce these holes. The winds are so cold that they super-cool the exposed surface waters. This leads to more sea-ice formation and salt rejection because when ice is formed it rejects any salt that the freezing water contains, which produces the coldest and saltiest water in the world. AABW flows around the Antarctic and penetrates the North Atlantic, flowing under the warmer and thus somewhat lighter NADW. The AABW also flows into both the Indian and Pacific Oceans. The NADW and AABW make up the key elements of the great global ocean conveyor belt (Figure 15), which allows heat to be exchanged

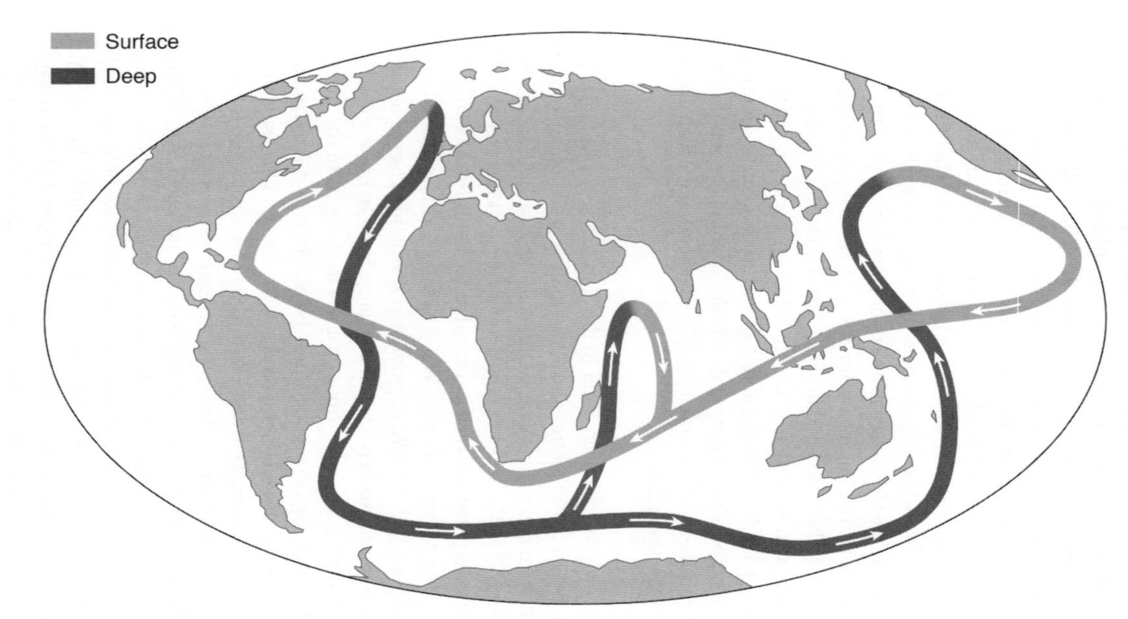

15. Global deep-ocean conveyor belt circulation

between the two hemispheres on the timescale of hundreds and thousands of years.

The balance between the NADW and AABW is extremely important in maintaining our present climate, as not only does it

Box 2 Blond hair and ocean circulation

The Gulf Stream may have also given us blond, fair skinned people. The warming effect of the Gulf Stream on Western Europe is so great that it means that early agriculturalists could grow crops incredibly far north in countries such as Norway and Sweden. These early settlers were living as far north as the Arctic Circle, which is on the same latitude as the middle of the Greenland ice sheet or the northern Alaska tundra. But there is one major drawback to living so far north and that is the lack of sunlight. Humans need Vitamin D, without it children develop rickets, which causes softening of the bones, leading to fractures and severe deformity. Vitamin D is made in the skin when it is exposed to ultraviolet light from the sun. This of course was no problem for our ancestors who evolved in Africa—quite the reverse, and dark skin was essential protection from the strong sunlight. However, as our ancestors moved further and further north there was less and less sunlight and less production of Vitamin D. In each generation only those with the lightest skin and hair colour could avoid getting rickets since the lighter your skin and hair, the more sunlight you can absorb, and thus the more Vitamin D you can make. So there was a very strong selection pressure in these areas in favour of fair skinned, blond haired people. On the other hand, Vitamin D is also found in food, such as fatty fish species and mushrooms, which may be why the same selection pressure did not apply to the Arctic Inuit. However, it is interesting to think that if it were not for the Gulf Stream and the stubbornness of the early Scandinavian settlers, relying only on crops and eating little or no fish, we would not have real blonds.

keep the Gulf Stream flowing past Europe, but it maintains the right amount of heat exchange between the Northern and Southern Hemispheres. Scientists are worried that the circulation of deep water could be weakened or 'switched off' if there is sufficient input of fresh water to make the surface water too light to sink. This evidence has come from both computer models and the study of past climates. Scientists have coined the phrase 'dedensification' to mean the removal of density by adding fresh water and/or warming up the water, both of which prevent seawater from being dense enough to sink. There is concern that climate change could cause parts of Greenland to melt. This could lead to more fresh water being added to the Nordic seas, thereby weakening the NADW and the Gulf Stream. This would bring much colder European winters with generally more severe weather. However, since the influence of the warm Gulf Stream is mainly in the winter, this change would not affect summer temperatures. So, if the Gulf Stream fails, global warming would still cause European summers to heat up. Europe would end up with extreme seasonal weather very similar to that of Alaska.

Global vegetation

The vegetation zones of the world are controlled by the annual average and seasonality of both temperature and precipitation. Temperature follows a latitudinal gradient with warmest conditions in the tropics and coldest at the poles. As we have seen there are three main rainfall belts, the convection rainfall belt in the tropics and the convergent rainfall belt in the mid-latitudes of the Northern and Southern Hemispheres. The two main desert regions of the world lie between these rainfall belts. Vegetation follows these climate zones. So rainforest is found in the tropics where there is a lot of rainfall all year round. Savannah is found in the tropics when rainfall is high seasonally, but there are also long dry seasons lasting over 4 months. The world's largest deserts are found in the mid-latitudes. Here the seasonality of rainfall is critical, as while many deserts have the same rainfall as, say,

London, this rain falls over a very small period of time, with the rest of the year being extremely arid. When the rainfall occurs only in the winter months followed by a very dry summer period, the unique Mediterranean flora is found, such as in California, South Africa, and of course around the Mediterranean. In high mid-latitudes are the temperate or boreal forests. In areas with low annual rainfall, steppe vegetation is found. In high latitudes where the temperature is the limiting factor, tundra is found. Other factors can influence where different vegetation can exist, for example we have seen that major ocean currents can allow temperate-weather vegetation to exist much further north than would usually be expected. In Chapter 5 we will see that mountain ranges and plateaus have a huge influence on where rainfall occurs and thus where deserts form.

Finally, it should be remembered that vegetation has its own influence on climate. First, vegetation changes the albedo of any area, so tropical rainforests absorb much more solar radiation than does tundra. Second, vegetation is very good at recycling water and maintaining a moist atmosphere. For example, 50 per cent of all the rainfall in the Amazon Basin comes from water recycled by the trees, evaporating and creating new clouds.

Chapter 3
Weather versus climate

Introduction

Many people get weather and climate confused. This confusion is exacerbated when scientists are asked to predict climate 50 years from now when everyone knows they cannot predict the weather a few weeks ahead. So climate is generally defined as 'the average weather'. The original definition of climate was 'the average weather over 30 years', this has been changed because we now know that our climate is changing and significant changes have been seen every decade for the last 50 years. The chaotic nature of the weather can make it unpredictable beyond a few days, while understanding the climate and modelling climate change is much easier as you are dealing with long-term averages. A good comparison is that though it is impossible to predict at what age any particular person will die, we can say with a high degree of confidence that the average life expectancy of a person in a developed country is about 80. The other confusion is that people always remember extreme weather events and not the average weather. So for example everyone remembers the heatwaves in the UK in 2003 and the USA in 2012, or the floods in Pakistan and Australia in 2010. So our perception of weather is skewed by these events rather than by an appreciation of the average weather or climate.

Chaos theory

The National Weather Service in the USA spends over $1 billion per year ensuring the country has the most accurate weather prediction possible. In other countries similar resources are poured into weather agencies, as predicting the weather is big business and getting a storm prediction right can save billions of dollars and many lives. Today three–four day forecasts are as accurate as the two-day forecasts were 20 years ago. Predictions of rain in three days' time are as accurate as one-day forecasts were in the mid-1980s. The accuracy of flash flood forecasts has improved from 60 per cent correct to 86 per cent, moreover potential victims of these floods get nearly an hour's warning instead of the 8 minutes they would have had in 1986. The lead times of advance warnings of tornadoes, in other words, the time that residents have to react, has increased from 5 minutes in 1986 to over 12 minutes. Severe local thunderstorms and similar cloudbursts are typically seen 18 minutes beforehand rather than 12 minutes over two decades ago. Seventy per cent of all hurricane paths can be predicted at least 24 hours in advance and the landfall of a hurricane can be predicted to within 100 miles (160 km).

These are great achievements but it does not explain why with all our technology and our understanding of the climate system we cannot predict the weather 10 days, a month, or a year in advance. Moreover, think of all those times that the weather report on television has said it will be sunny today and then it rains. So why is it so difficult to predict the weather? In the 1950s and 1960s it was thought that our weather prediction was limited by our lack of data and that if we could measure things more accurately and clearly understand the fundamental processes we would be able to achieve a much higher level of prediction. But in 1961 Edward Lorenz a meteorologist at Massachusetts Institute of Technology made a cup of coffee that radically changed the way we think about natural systems. In 1960 Lorenz had produced one of the first computer models of weather. One day in the winter of 1961, Lorenz's computer model produced some very interesting patterns, which he wanted to look at in greater detail. So he took a short-cut and started mid-way through the run. Of course this was one of the earliest computers so he had to retype all the starting numbers. Instead of typing them into six decimal places (e.g., 0.506127) he only typed the first three to save time and space, and then went and made the famous coffee. When Lorenz came back he found that the weather patterns had diverged from the initial run so much that there was no recognizable similarity between them. It seems the model was very sensitive to the very small changes, that one part in a thousand instead of being inconsequential had had a huge effect on the outcome. This original work has lead to the development of chaos theory. Chaos theory shows us that very small variations in atmosphere temperature, pressure, and humidity can have a major and unpredictable or chaotic effect on large-scale weather patterns.

Nevertheless, chaos theory does not mean there is a complete lack of order within a system. Far from it, chaos theory tells us that we can predict within certain boundaries what the weather will be like: we all know, for example, that most tornadoes occur in May in the USA and that winters are wet in England. But when it comes to more detailed prediction everything breaks down due to what has become known as the 'butterfly effect'. The idea is that small changes represented by the flapping of the wings of a butterfly can have a large effect on the weather, for example altering the strength and direction of a hurricane. As errors and uncertainties multiply and cascade upward through the chain of turbulent features from dust devils and squalls up to continental size eddies that only satellites can see. In effect, we will never know which of the small weather changes will combine to have these large effects. While Lorenz used 12 equations in his weather model, modern weather computers use 500,000. But even the best forecasts, which come from the European Centre for Medium Weather Forecasts based at Reading in England, suggest that weather predictions for more than four days are at best speculative and beyond a week worthless, all because of chaos.

So chaos theory says that we can understand weather and we can predict general changes but it is very difficult to predict individual events such as rain storms and heatwaves. The study of climate, however, has one great advantage over meteorology because it only examines averages and thus chaos theory does not affect it. Moreover when it comes to modelling future climate change we can now understand that an increase in the Earth's average temperature will make some weather phenomena more frequent and intense for example heatwaves and heavy rainfall events, while others will become less frequent and intense, for example extreme cold events and snow fall.

Decadal and quasi-periodic climate systems

The climate system contains many cycles and oscillations that complicate our ability to predict the weather. These include decadal cycles such as the North Atlantic oscillation (NAO), the Atlantic multi-decadal oscillation (AMO), Arctic oscillation (AO), and the Pacific decadal oscillation (PDO). So the first of these is the NAO, which was first described in the 1920s by Sir Gilbert Walker (14 June 1868–4 November 1958), a British physicist and statistician. The NAO is a climate phenomenon in the North Atlantic Ocean and is represented by the atmospheric pressure difference at sea level between Iceland and the Azores. The difference in Icelandic low-pressure and the Azores high-pressure systems seems to control the strength and direction of westerly winds and storm tracks across the North Atlantic Ocean. This in turn controls where and when in Europe it rains. Unlike the El Niño–Southern Oscillation, the NAO is largely controlled by changes in the atmosphere. The NAO is closely related to the AO and though both seem to change on a decadal scale there seems to be no periodicity. The NAO should not, however, be confused with the AMO.

The AMO is the decadal-scale variability in the sea-surface temperatures of the North Atlantic Ocean. Over the last 130 years, 1885–1900, 1927–47, 1951–61, 1998–present day, the North Atlantic Ocean temperatures have been warmer than average and the time in between colder. The AMO does affect air temperatures and rainfall over much of the Northern Hemisphere, in particular North America and Europe, for example the North Eastern Brazilian and African Sahel rainfall and North American and European summer climates. It is also associated with changes in the frequency of North American droughts and it may influence the frequency of severe Atlantic hurricanes. There are also irregular or quasi-periodic cycles such as the Indian Ocean Dipole and El Niño–Southern Oscillation (ENSO). Of these ENSO is by far the best known and is discussed in more detail below.

El Niño–Southern Oscillation

One of the most important and mysterious elements in global climate is the periodic switching of direction and intensity of ocean currents and winds in the Pacific. Originally known as El Niño ('Christ child' in Spanish) as it usually appears at Christmas, and now more often referred to as ENSO (El Niño–Southern Oscillation), this phenomenon typically occurs every 3 to 7 years. It may last from several months to more than a year. ENSO is an oscillation between three climates: the 'normal' conditions, La Niña, and 'El Niño'. ENSO has been linked to changes in the monsoon, storm patterns, and occurrence of droughts throughout the world. For example the prolonged ENSO event, in 1997 to 1998, caused severe climate changes all over the Earth including droughts in East Africa, northern India, north-east Brazil, Australia, Indonesia, and Southern USA; and heavy rains in California, parts of South America, the Pacific, Sri Lanka, and east central Africa. The state of the ENSO has also been linked into the position and occurrence of hurricanes in the Atlantic Ocean. For example, it is thought that the poor prediction of where Hurricane Mitch (see Chapter 4) made landfall was because the ENSO conditions were not considered and the strong Trade Winds helped drag the storm south across central USA instead of west as predicted.

An El Niño event is when the warm surface water in the western Pacific moves eastward across to the centre of the Pacific Ocean (Figure 16). Hence the strong convection cell or warm column of rising air is much closer to South America. Consequently the Trade Winds are much weaker and the ocean currents crossing the Pacific Ocean are weakened. This reduces the amount of the cold, nutrient-rich upwelling off the coast of South America and without those nutrients the amount of life in the ocean is reduced and fish catches are dramatically reduced. This massive shift in ocean currents and the position of the rising warm air changes the direction of the jet streams that upset the weather in North America, Africa, and the rest of the world. However if you ask what causes El Niño, then the answer is of the chicken and egg variety. Does the westward ocean current across the Pacific reduce in strength, allowing the warm pool to spread eastward and moving with it the wind system. Or does the wind system relax in strength, reducing the ocean currents, and allowing the warm pool to move eastwards? Many scientists believe that long period waves in the Pacific Ocean that move between South America and Australia over time help shift the ocean currents which produce either an El Niño or a La Niña period.

La Niña is a more extreme version of the 'normal' conditions. Under normal conditions the Pacific warm pool is in the western Pacific and there are strong westerly winds and ocean currents keeping it there. This results in upwelling off the coast of South America, providing lots of nutrients and thus creating excellent conditions for fishing. During a La Niña period the temperature difference between the western and eastern Pacific becomes extreme and the westerly winds and ocean currents are enhanced. La Niña impacts on the world's weather are less predictable than those of El Niño. This is because during an El Niño period the

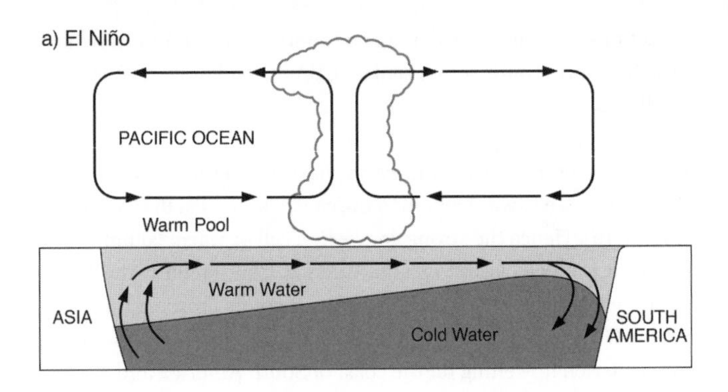

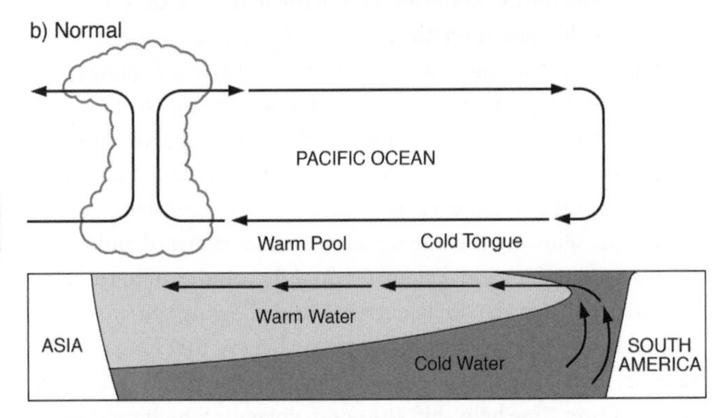

16. Pacific Ocean during El Niño and normal periods

Pacific jet stream and storm tracks become strong and straighter and it is therefore easier to predict its effects. La Niña on the other hand weakens the jet stream and storm tracks, making them more loopy and irregular, meaning that the behaviour of the atmosphere and particularly of storms becomes more difficult to predict. In general where El Niño is warm, La Niña is cool, where El Niño is wet, La Niña is dry. La Niñas have occurred in 1904, 1908, 1910, 1916, 1924, 1928, 1938, 1950, 1955, 1964, 1970, 1973, 1975, 1988, 1995, 1999, 2008, and 2011, with the 2010–11 La Niña being one of the strongest ever observed.

Predicting ENSO

Predicting an El Niño events is difficult but a lot of work has gone on for the last three decades to better understand the climate system. For example, there is now a large network of both ocean and satellite monitoring systems over the Pacific Ocean, primarily aimed at recording sea-surface temperature, which is the major indicator of the state of the ENSO. By using this climatic data in both computer circulation models and statistical models, predictions are made of the likelihood of an El Niño or La Niña event. We are really still in the infancy stage of developing our understanding and predictive capabilities of the ENSO phenomenon.

There is also considerable debate over whether ENSO has been affected by global warming. The El Niño conditions generally occur every 3 to 7 years; however, in the last 20 years, they have behaved very strangely, returning for 3 years out of 4: 1991–2, 1993–4, and 1994–5, then not returning until 1997–8, and then not returning for 9 years, finally arriving in 2006–7. Reconstruction of past climate using coral reefs in the western Pacific shows sea-surface temperature variations dating back 150 years, well beyond our historical records. The sea-surface temperature shows the shifts in ocean current, which accompany shifts in the ENSO and reveal that there have been two major changes in the frequency and intensity of El Niño events. First was a shift at the beginning of the 20th century from a 10–15-year cycle to a 3–5-year cycle. The second was a sharp threshold in 1976 when a marked shift to more intense and even more frequent El Niño events occurred. Moreover during the last few decades the number of El Niño events has increased, and the number of La Niña events has decreased. Even taking into account the effect of decadal cycles on ENSO the size of the ENSO variability in the observed data seems to have increased by 60 per cent in the last 50 years.

However, as we have seen, to predict an El Niño event 6 months from now is hard enough, without trying to assess whether or not

ENSO is going to become more extreme over the next 100 years. Most computer models of ENSO in the future are inconclusive; some have found an increase and others have found no change. This is, therefore, one part of the climate system that we do not know how global warming will affect. Not only does ENSO have a direct impact on global climate but it also affects the numbers, intensity, and pathways of hurricanes and cyclones, and the strength and timing of the Asian monsoon. Hence, when modelling the potential impacts of global warming, one of the largest unknowns is the variation of ENSO and its knock-on effects on the rest of the global climate system.

Modelling climate

The whole of human society operates on knowing the future weather. For example, a farmer in India knows when the monsoon rains will come next year and so they know when to plant the crops. While a farmer in Indonesia knows there are two monsoon rains each year so each year they can have two harvests. This is based on their knowledge of the past, as the monsoons have always come at about the same time each year in living memory. But weather prediction goes deeper than this as it influences every part of our lives. Our houses, roads, railways, airports, offices, cars, trains, and so on are all designed for our local climate. Predicting future climate is, therefore, essential as we know that global warming is changing the rules. This means that the past weather of an area cannot be relied upon to tell you what the weather in the future will hold. So we have to develop new ways of predicting and modelling the future, so that we can plan our lives and so that human society can continue to fully function.

There is a whole hierarchy of climate models, from relatively simple box models to the extremely complex three-dimensional general circulation models (GCMs). Each has a role in examining and furthering our understanding of the global climate system. However, it is the complex three-dimensional general circulation models which are used to predict future global climate. These comprehensive climate models are based on physical laws represented by mathematical equations that are solved using a three-dimensional grid over the globe. To obtain the most realistic simulations, all the major parts of the climate system must be represented in sub-models, including the atmosphere, ocean, land surface (topography), cryosphere, and biosphere, as well as the processes that go on within them and between them. Most global climate models have at least some representation of each of these components. Models that couple together both the ocean and atmosphere components are called atmosphere–ocean general circulation models (AOGCMs).

Over the last 25 years there has been a huge improvement in climate models. This has been due to our increased knowledge of the climate system but also because of the nearly exponential growth in computer power. There has been a massive improvement in spatial resolution of the models from the very first Intergovernmental Panel on Climate Change (IPCC) in 1990 to the latest in 2007. The current generation of AOGCMs has a resolution of one point every 110 km by 110 km, and this is set to get even finer when the next IPCC Science Report is published in late 2013. The very latest models or as some groups are now referring to them 'climate simulators' include much better representations of atmospheric chemistry, clouds, aerosol processes, and the carbon cycle including land vegetation feedbacks. But the biggest unknown or error in the models, is not the physics, it is the estimation of future global greenhouse emissions over the next 90 years. This includes many variables, such as the global economy, global and regional population growth, development of technology, energy use and intensity, political agreements, and personal lifestyles.

Over 20 completely independent AOGCMs have been run using selected future carbon dioxide emission scenarios for the IPCC 2007 report, producing global average temperature changes that may occur by 2100. This is a significant change from the IPCC 2001 report, in which only 7 of these models were used. Using the widest range of potential emission scenarios the climate models suggest that global mean surface temperature could rise by between 1.1°C and 6.4°C by 2100. Using the best estimates for the 6 most likely emission scenarios, then this range is 1.8°C to 4°C by 2100. Model experiments show that even if all radiation forcing agents were held at a year-2000 constant, there would still be an increase of 0.1°C per decade over the next 20 years. This is mainly due to the slow response of the ocean. Interestingly, the choice of emission scenario has little effect on the temperature rise to 2030, making this a very robust estimate. All models suggest twice the rate of temperature increase in the next two decades compared with that of the 20th century. What is significant is that the choices we make now in terms of global emissions will have a significant effect on global warming after 2030. The next IPCC report to be published in late 2013, though it will use greatly improved emission scenarios, will have a very similar potential change of warming by the end of the century. What is amazing and very reassuring is that over the last 25 years the climate models have consistently given us the same answer, meaning we do understand the climate system and we can understand the consequences of our past and future actions.

Chapter 4
Extreme climates

Introduction

Humans can live, survive, and even flourish in the extreme climates ranging from that of the Arctic to that of the Sahara. We have populated every continent except Antarctica. We can deal with the average climate of each region through our adaptations of technology and lifestyle. The problems arise when the predictable boundaries of local climate are exceeded, for example by heatwaves, storms, droughts, and/or floods. This means that what we define as extreme weather, such as a heatwave, in one region may be considered fairly normal weather in another. Each society has a coping range, a range of weather with which it can deal: what is seen as a heatwave in England would be normal summer conditions in Kenya. However, one of the most unpredictable and dangerous elements in our climate systems are storms. In this chapter we examine how and why storms are formed and their impact. Hurricanes, tornadoes, winter storms, and the monsoons will all be discussed.

Hurricanes

A hurricane is a severe cyclonic tropical storm that starts in the North Atlantic Ocean, Caribbean Sea, Gulf of Mexico, west coast of Mexico, or the northeast Pacific Ocean (Figure 17). They are called typhoons in the western Pacific and simply tropical cyclones in the Indian Ocean and Australasia. They are, however, all the exactly the same type of storm and here we will call all of them hurricanes. Hurricanes occur in the tropics between 30°N and 30°S, but not near the Equator as there is not enough atmospheric variation to generate them. For a storm to be classified as a hurricane, the sustained wind speed must exceed 120 km/hr. Of course in a fully developed hurricane, wind speeds can exceed 200 km/hr.

A hurricane is a tropical storm run amok, a rotating mass of thunderstorms that has become highly organized into circular cells, which are ventilated by bands of roaring winds. Hurricanes develop over the oceans and tend to lose their force once they move over land—this is because unlike temperate storms hurricanes are driven by the latent heat from the condensation of water. The sun is most intense close to the Equator where it heats the land, which in turn heats the air. This hot air rises and consequently sucks air from both the north and south producing the Trade Winds. As the seasons change so does the position of the clash of the Trade Winds, which is called the Inter-tropical Convergence Zone (ITCZ). To generate a hurricane the sea temperature must be above 26°C for at least 60 m below the surface and the air humidity must be at about 75–80 per cent. This combination provides the right amount of heat and water vapour to sustain the storm once it has started. For example these conditions occur during the summer in the North Hemisphere when the tropical North Atlantic Ocean heats up enough and its water starts to evaporate. Initially the warm ocean heats the air above it and causes that to rise. This produces a low-pressure area which sucks in air from the surrounding area. This rising air contains a lot of water vapour due to pronounced evaporation from the hot surface of the ocean. As the air rises it cools and can no longer hold as much water vapour; as a result some of it condenses to form water droplets and then clouds. This transformation from water vapour to water droplets releases energy called 'latent heat'. This in turn causes further warming of

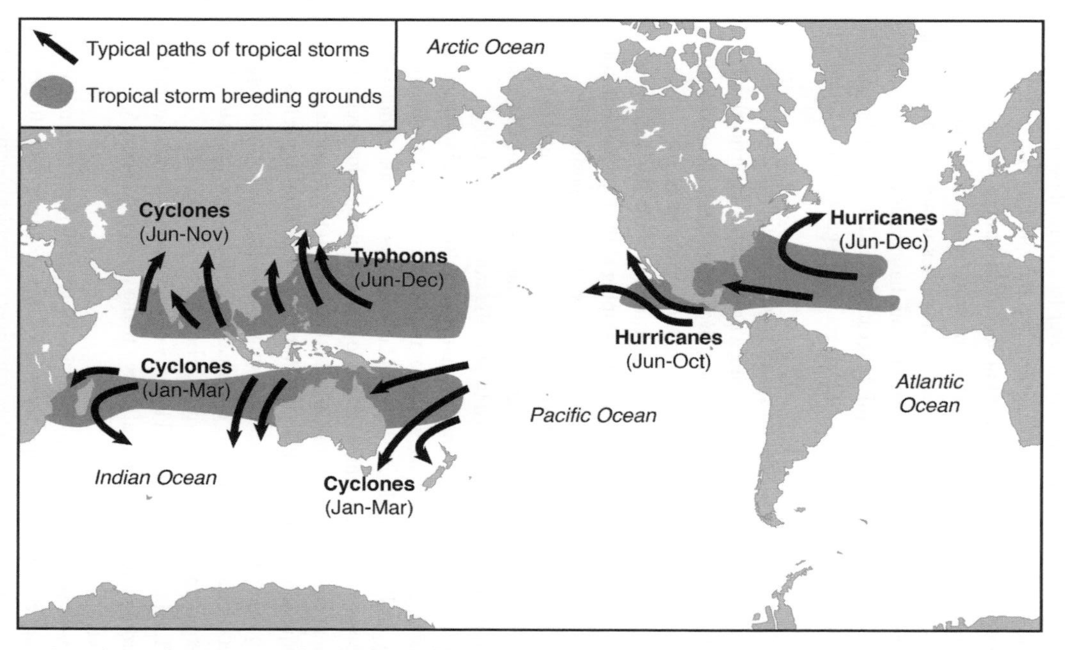

17. Location and occurrence of major tropical storms

the air and causes it to rise even higher. This feedback can make the air within a hurricane rise to over 10,000 m above the ocean. This becomes the eye of the storm and the spiraling rising air it produces creates a huge column of cumulo-nimbus clouds. You can see a mini version of this with steam coming out of a kettle. As the hot air rises from the kettle it hits the colder air and it forms steam, a mini-cloud. If you have ever put your hand near the steam you can feel it is very hot and this is because of all the energy being released as the water vapour changes from a gas back to a liquid.

When the air inside the hurricane reaches its highest level it flows outwards from the eye producing a broad canopy of cirrus cloud. The air cools and falls back to sea level where it is sucked back into the centre of the storm. Because of the Coriolis force, the air that is sucked into the bottom of the hurricane spins into the storm in a clockwise direction, while the air escaping at the top spins out in a counter-clockwise direction. This pattern is the opposite in the Southern Hemisphere. Hurricanes form at least 345 miles or 5° of latitude away from the Equator, where the Coriolis effect is strong enough to give the required twist to the storm. The size of hurricanes can vary from 100 km to over 1,500 km. A hurricane can form gradually over a few days or in the space of 6 to 12 hours and typically the hurricane stage will last 2–3 days and take about 4–5 days to die out. Scientists estimate that a tropical cyclone releases heat energy at the rate of 50 to 200 exajoules (10^{18} J) per day, equivalent to about 1 PW (10^{15} Watt). This rate of energy release is equivalent to 70 times the human world energy consumption and 200 times the worldwide electrical generating capacity, or to exploding a 10-megaton nuclear bomb every 20 minutes. Hurricanes are measured in the Saffir-Simpson scale and go from a tropical storm through category 1 to the worst at category 5.

However, the formation of hurricanes is much rarer than might be expected given the opportunities for them to occur. Only 10 per cent of centres of falling pressure over the tropical oceans give rise to fully fledged hurricanes. In a year of high incidence, perhaps a maximum of 50 tropical storms will develop to hurricane levels. Predicting the level of a disaster is difficult as the number of hurricanes does not matter—it is whether they make landfall (Figure 18). For example, 1992 was a very quiet year for hurricanes in the North Atlantic Ocean. However, in August, one of the few hurricanes that year, Hurricane Andrew, hit the USA just south of Miami and caused damage estimated at $26 billion. Hurricane Andrew also demonstrates that predicting where a storm will hit is equally important—if the hurricane had hit just 20 miles further north it would have hit the densely populated area of Miami City and the cost of the damage would have doubled.

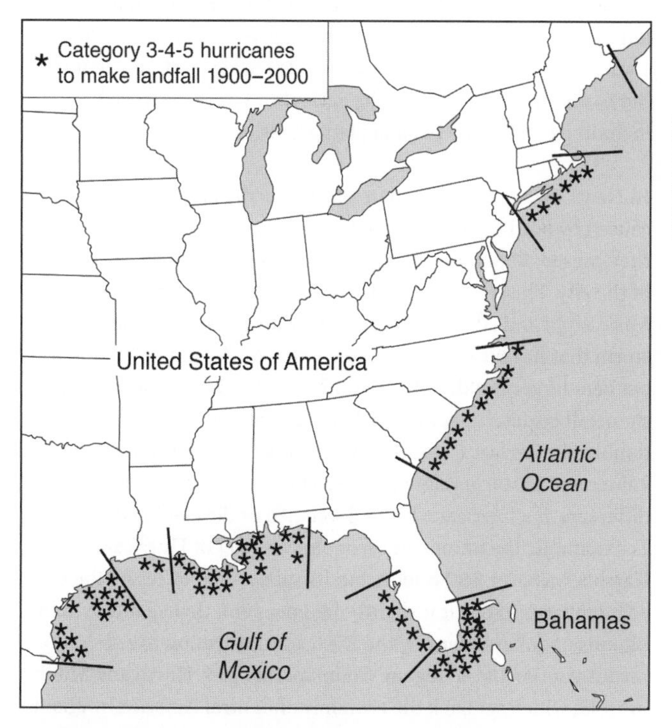

18. Major hurricane landfall over a 100-year period

In terms of where hurricanes hit in developed countries, the major effect is usually economic loss, while in developing countries the main effect is loss of life. For example, Hurricane Katrina, which hit New Orleans in 2005, caused 1,836 deaths while Hurricane Mitch, which hit Central America in 1998, killed at least 25,000 people and made 2 million others homeless. In both cases the greatest damage was caused by the huge amount of rainfall. Honduras, Nicaragua, El Salvador, and Guatemala were battered by 180-mile (290 km) per hour winds, and more than 23 inches (60 cm) of rain every day. Honduras, a small country of only 6 million inhabitants, was the worst hit. The Hamuya River, normally a calm stretch of water about 200 feet (60m) wide rose by 30 feet (9 m) and became a raging torrent, ripping out trees as tall as a city block from the ground. Eighty-five per cent of the country ended up under water. Over 100 bridges, 80 per cent of the roads, and 75 per cent of its agriculture were destroyed, including most of the banana plantations.

In New Orleans the worst damage by Hurricane Katrina was caused by both the intense rainfall and the storm surge. Together they caused 53 different levees to break, submerging 80 per cent of the city. The storm surge also devastated the coasts of Mississippi and Alabama. Hurricane Katrina was not the worst storm that has hit the USA; a storm that hit Miami in 1926 was 50 per cent larger but did little damage because Miami Beach was then still undeveloped. In the USA coastal population has doubled in the last 10 to 15 years making the country much more vulnerable to storm related losses. There is also a large financial difference if a hurricane hits a developed or developing country. For example, the immediate economic impact of Hurricane Katrina was over $80 billion, but its subsequent effect on the US economy was to boost it slightly, by 1 per cent, that year due to the billions of dollars spent by the Bush administration to aid reconstruction of the region. Compare this with Hurricane Mitch in 1998, which set back the economy of Central America by about a decade.

Hurricanes also occur elsewhere in the world. An average of 31 tropical storms roam the western North Pacific every year, with typhoons smashing into Southeast Asia from June to December; most at risk are Indonesia, Hong Kong, China, and Japan, otherwise known as 'Typhoon Alley'. Why does Typhoon Alley get so many typhoons? And why can they occur almost any time of year. The answers lie in the oceans. The key is the 'warm pool' of ocean water that sits in the western tropical Pacific. All year long the Trade Winds and the ocean current push the surface water warmed by the tropical sun to the far western side of the North Pacific. Hurricane seasons come and go in other parts of the world but the water of the 'warm pool' is always warm enough to start a hurricane—though they are most common between June and December (Figure 18).

Tornadoes

Tornadoes are nature's most violent storms. Nothing that the atmosphere can dish out is more destructive: they can sweep up anything that moves; and they can lift buildings from their foundations, making a swirling cloud of violently flying debris. They are very dangerous, not only because of the sheer power of their wind, and the missiles of debris they carry, but because of their shear unpredictability. Tornado strength and destructive capability is measured on the Fujita Scale.

A tornado is a violent rotating column of air, which at a distance appears as an ice cream cone-shaped cloud formation. Other storms similar to tornadoes in nature are whirlwinds, dust-devils (weaker cousins of tornadoes occurring in dry lands), and waterspouts (a tornado occurring over water). Tornadoes are most numerous and devastating in central, eastern, and northeastern USA, where an average of 5 per day are reported every May. They are also common in Australia (15 per year), Great Britain, Italy, Japan, Bangladesh, east India, and central Asia. While the greatest number of fatalities occurs in the United States, the deadliest tornadoes by far have occurred in the small area of Bangladesh and east India. In this 8,000 mile^2 (21,000 km^2) area, 24 of the 42 tornadoes known to have killed more than a 100 people have occurred. This is likely due to the high population density and poor economic status of the area as well as a lack of early warning systems.

We can see tornadoes as miniature hurricanes. Although tornadoes can form over tropical oceans they are more common over land. The formation of tornadoes is encouraged when there is warm, moist air near the ground and cold dry air above. This occurs frequently in late spring and early summer over the Great Plains of the USA (Figure 19). Intense heating of the ground by the sun makes warm, moist air rise. As it does so it cools and forms large cumulo-nimbus clouds. The strength of the updraft determines how much of the surrounding air is sucked into the bottom of what becomes a tornado. Two things help the tornado to rotate violently; the first is the Coriolis force and the second is the high level jet stream passing over the top of the storm, adding an extra twist to the tornado. Because of the conditions under which tornadoes are formed they can easily occur beneath thunderstorms and hurricanes.

In the USA nearly 90 per cent of tornadoes travel from the southwest to the northeast, although some follow quick changing zigzag paths. Weak tornadoes or decaying tornadoes have a thin ropelike appearance. The most violent tornadoes have a broad dark funnel shape that extends from the dark wall cloud of a large thunderstorm. There have even been reports of some tornadoes practically standing still, hovering over a single field, and of others that crawl along at 5 miles per hour. On the other hand, some have been clocked at over 70 miles per hour. However, on average, tornadoes travel at 35 miles per hour. It has been noted that most tornadoes occur between 3pm and 9pm, but they have been known to strike at any time of day or night. They usually only last about 15 minutes, staying only a matter of seconds in any single place—but

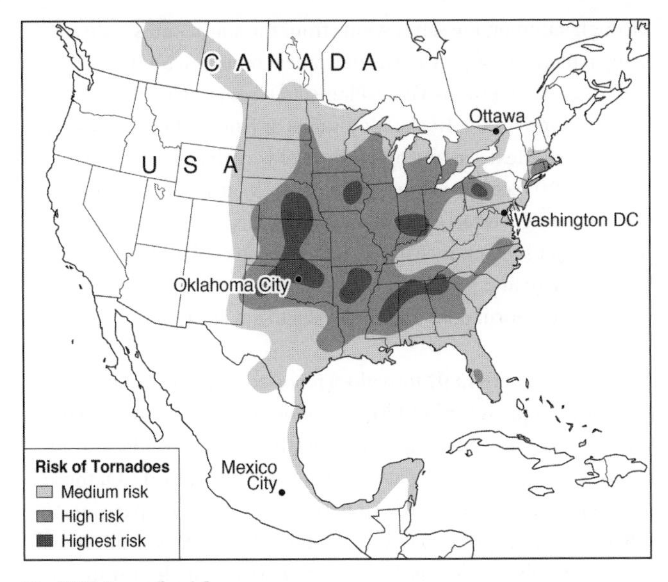

19. USA tornado risk map

then some tornadoes just do not fit any of these rules, for example on 18 March 1925 a single tornado travelled 219 miles in 3.5 hours through Missouri, Illinois, and Indiana killing 695 people.

Tornado Alley

Tornado Alley is the nickname for the area in which most tornadoes occur in the USA, and it expands through spring and summer as the heat from the sun grows warmer and the flow of warm moist air from the Gulf of Mexico spreads further north (see Figure 19). An area that includes central Texas, Oklahoma, and Kansas is at the hard core of Tornado Alley, but before the season is over it can have expanded to the north to Nebraska and Iowa. It shrinks and swells over time but there is only one Tornado Alley. Nowhere else in the world sees weather conditions in a combination so perfect to make tornadoes. The key reasons for this special area are: (1) beginning in spring and continuing through summer, low-level winds from the south and southeast bring a plentiful supply of warm tropical moisture up from the Gulf of Mexico into the Great Plains; (2) from down off the eastern slopes of the Rocky Mountains or from out of the deserts of northern Mexico come other flows of very dry air that travel about 3,000 feet above the ground; and (3) at 10,000 feet high the prevailing westerly winds, sometimes accompanied by a powerful jet stream, race overhead, carrying cool air from the Pacific Ocean and providing a large temperature difference, which will drive the tornadoes and the twists to get started.

In 2011 there were 1,897 tornadoes reported in Tornado Alley in the USA beating the record of 1,817 tornadoes recorded in 2004. The year 2011 was also an exceptionally destructive and deadly year in terms of tornadoes, killing at least 577 people worldwide. Of those, an estimated 553 were in the United States, which compared to 564 US deaths in the prior 10 years combined. That year saw the second greatest number of deaths due to tornadoes in a single year in US history. However, this is still a long way off from the most deadly tornado on record, which occurred on 26 April 1989 in Bangladesh and killed over 1,300 people, injured 12,000 people, and destroyed everything but a few trees from Daultipur to Salturia.

Winter storms

For people living in the mid-latitudes weather seems to be a permanent topic of conversation. This is because it is always changing. In Britain there is a saying, 'if you do not like the weather wait an hour and it will change'. This is because the climate of the mid-latitudes is dominated by the titanic clash between the cold polar air moving southward and the warm sub-tropical air moving northwards. This clash of air masses takes place at the Polar Front.

The Polar Front moves north and south with the seasons. In summer when the sub-tropical air is warmer it moves further towards the pole. During winter when conditions are much colder the polar air mass is dominant and the Polar Front moves towards the Equator. Where these two air masses meet rain is formed. This is because warm air can hold more water vapour and when it clashes with the cold air this vapour condenses into clouds, which in turn produce rain. But it is the upper atmosphere which really controls the shape and thus the weather of the Polar Front. The upper atmosphere is characterized by fast 'jet streams' that race around the planet. These powerful jet streams push the Polar Front around the Earth, but as it does so the Front wrinkles and becomes a mass of so-called planetary waves moving gradually round our planet. These waves have a great effect on our weather, causing us all to complain about the weather being so changeable and of course wet. One of these waves can pass over a town in about 24 hours. The weather will be experienced as starting out to be relatively cold but with clear skies. As the warm front passes overhead the conditions get warmer and it starts to rain—usually light rain or drizzle. As the centre of the warm air mass reaches the town the weather turns cloudy and muggy and the rain stops. Then the second front, the cold front, passes overhead; temperatures drop and there is a short period of very heavy rainfall. Then it is back to cold, clear weather until the next wave reaches the town.

As we have seen there are many storms that are associated with distinct areas of atmospheric circulation described in the section on Hadley Cells. Ice, wind, hail, and snow storms are associated with either the Polar Front or high mountain regions and are worse in the winter time. In the Northern Hemisphere these types of storms are common over North America, Europe, Asia, and Japan.

For snow to reach the ground the temperature of the air between the base of the cloud and the ground must be below 4°C, otherwise the snowflakes melt as they travel through the air. For hailstones to form the top of the storm must be very cold. High up in the atmosphere water droplets can become super cooled to less than 0 °C, which collide in the atmosphere to form ice balls or hailstones. If you cut open a hailstone you can see the layers of ice that have built-up like an onion. The stones can vary from between 2 mm and 20 cm. Their size depends on how strong the

Box 3 Caught in the cold

When your body loses the battle against the cold, it is often someone else who will notice it. This is why you should always be on the look-out for the symptoms of cold weather exposure in your companions. When the cold has started to affect you badly, you are not always the best judge of the seriousness of the problem. You still think that you are okay you just need another minute's rest. These are the signs to look out for:

- You cannot stop shivering
- You are fumbling your hands
- Your speech is slow and slurred and may even be incoherent
- You stumble and lurch as you walk
- You are drowsy and exhausted and feel the need to lie down even though you are outside
- Maybe you have rested, but cannot then get up

A person acting like this needs to get into dry clothes and a warm bed. This is because the core temperature of that person has started to drop, which is extremely dangerous for the body; if it is not stopped it will result in death. They need a warm hot water bottle, heating pad, or warm towels on their body. They need warm drinks. They do NOT need an alcoholic or caffeinated drink, as these speed up the person's heart rate, causing them to lose yet more heat; they also dehydrate the body, which hinders its recovery. Also do NOT massage or rub the person, as this again takes away heat from the body core where it is most required. The person should also always be seen by a doctor.

updraft of air is, as this determines how long they stay in the atmosphere before dropping out. The worst storm conditions are called blizzards. These combine strong winds, driving snow, ice, and hail, with air temperatures as low as −12 °C and visibility less than 150 metres (see Box 3).

Monsoons

The other important area for massive rainstorms is the monsoon belt. The name monsoon comes from the Arabic word 'mausim' which means 'season', as most of the rains that fall in Southeast Asia occur during the summer. In the tropics the sun's energy is most intense as the sun is directly overhead. This heats up the land and sea and thus warms the air above. This warm, moist air rises, leaving an area of low pressure beneath it, which helps to suck in air from the surrounding area (Figure 20). This suction results in the Trade Winds, which can travel from much higher latitudes to this area of rising air. As the winds come from both the Northern and Southern Hemispheres this area is known as the ITCZ. As the air at the ITCZ rises, it forms huge towering clouds and produces large amounts of rain. The ITCZ moves north and south with the seasons as the position of the most intense sunlight shifts up and down across the Equator. It is also influenced strongly by the position of the continents. This is because the land heats up faster and to a greater extent than the ocean and thus it can pull the ITCZ even further north or south during that season. An example of this is the Asian summer monsoon, during the summer near the Himalayan mountains and the low lands of India heat up. This pulls the ITCZ across the Equator on to Asia. Because the Southern Hemisphere winds have been pulled across the warm Indian Ocean they are warm and full of moisture; when they are forced to rise and cool down over India they produce very heavy rainfall throughout Southeast Asia and as far north as Japan. During Northern Hemisphere winter the ITCZ moves south of the Equator, but in Southeast Asia it means warm, moist winds from the North Pacific are dragged southward across the

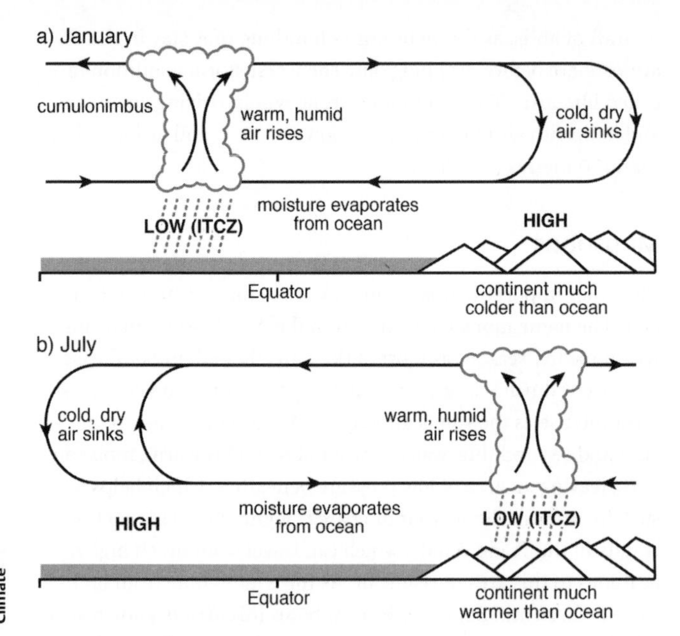

20. Monsoonal system

continent into the Southern Hemisphere. This means that some areas such as Indonesia and Southern China get two monsoonal rainy seasons a year: one from the north and one from the south. No wonder this is the most fertile place on Earth, supporting over two-fifths of the world's population. Despite being the bringers of life, the rains can cause catastrophic hazards, especially in the form of flooding. Examples of this are the terrible floods in 1998 in Bangladesh and China which caused over $30 billion of damage and thousands of deaths.

Amazon monsoon

During the Southern Hemisphere summer the continent of South America heats up. This rising air leaves an area of low pressure at ground level, which is filled by sucking in the surrounding air.

This pulls the convergence zone between the North and Southern tropical air southward over Brazil. The southward shift of the ITCZ brings with it lots of rain as the air being pulled across the Equator from the north originates over the warm, tropical Atlantic Ocean. This produces the Amazon monsoon and results in the mightiest river in the world and the greatest extent of rainforest on the planet. The Amazon Basin covers an amazing 2.7 million $miles^2$ much of which is covered with rainforest. The Amazon River delivers 20 per cent of all the freshwater that enters the world's oceans. Without the monsoon rains the most diverse habitat in the world would not exist.

Living under the Asian monsoons

Bangladesh is a country literally built by the monsoons as over three-quarters of the country is a deltaic region formed by the sediments brought in by the Ganges, Brahmaputra, and Meghna rivers—all fed by the summer monsoons. Over half the country lies less than 5 metres above sea level, thus flooding is a common occurrence. During a normal summer monsoon a quarter of the country is flooded. Yet these floods, like those of the Nile, bring life with them as well as destruction. The water irrigates and the silt fertilizes the land. The fertile Bengal Delta supports one of the world's most dense populations, over 110 million people in 140,000 km^2. But every so often the monsoon floods exceed what even Bangladesh can cope with. In 1998 three-quarters of the country was flooded for 2 months, causing billions of pounds worth of damage and thousands of deaths. Bangladesh also has to cope with tropical cyclones. If we take 3 of the worst years for tropical cyclones we can see the loss of life has dramatically dropped. In 1970 there were over 300,000 cyclone-related deaths, in 1991 there were 138,000, while in 2007 there were just 3,500 deaths. This is not because the tropical cyclones have grown gentler, far from it. It is because of good governance. The Bangladesh government has, first, invested in excellent meteorological facilities to make as accurate a prediction of when and where the cyclones will make land fall; second, they have set up a communication network using cyclists, so that once a cyclone warning is given, the message is carried to all the towns and villages that will be affected. They have also built cyclone shelters, protected water and sanitation facilities, and encouraged floating agriculture, which can withstand the storms. These relatively simple changes have resulted in the saving of hundreds of thousands of lives.

Chapter 5
Tectonics and climate

Introduction

In Chapters 1 and 2 we saw how climate is a function of how the sun's energy falls on the Earth and is then redistributed around the globe. Both of these aspects are strongly influenced by plate tectonics. This is why 100 million years ago the Earth was much warmer and humid, and dinosaurs were happily living on Antarctica. Our modern climate system is a product of millions of years of plate tectonics, which have produced unique occurrences such as significant amounts of ice at both Poles. This produces a very strong Equator-pole temperature gradient and thus an extremely dynamic and energetic climate system. Tectonics has two main effects on climate. First, there are direct effects, which include mountain and plateau uplift which changes atmospheric circulation and the hydrological cycle or ocean gateways, which change the way the oceans circulate. Second, there are indirect effects that affect the content of the atmosphere through subduction, volcanism, and consumption of gases by chemical weathering. One of the themes running through this book is the idea that nothing in climatology is complex. This is also true of the effects of tectonics on climate. In this chapter the influences are broken down into horizontal tectonics, which examines what happens if you simply move the continental plates around the globe. Next is vertical tectonics, which examines what happens if you create a mountain or a plateau. Last, we will look at the effects of volcanoes and supervolcanoes on climate.

Horizontal tectonics

Latitudinal continents

The north–south position of the continents has a huge effect on the thermal gradient between the poles and the Equator. Geologists have run simple climate models to look at this effect (Figure 21). If you put all the continents around the Equator, the so-called 'tropical ring world', the temperature gradient between the poles and the Equator is about 30 °C (Figure 21). This is because when the poles are covered with oceans they cannot go below freezing. This is due to a trick of both the atmosphere and the oceans. A fundamental rule of climate is that hot air rises and cold air drops. At the poles it is cold so the air falls and as it hits the ground it pushes outwards away from the pole. When sea water at the pole freezes it forms sea ice, and this ice is then blown away from the pole towards warmer water where it melts. This maintains the balance and prevents the temperature of the poles going below zero. However, as soon as you introduce land onto the pole or even around the pole, ice can form permanently. If you do have a landmass like Antarctica over a pole with ice on it the Equator–pole temperature gradient is over 65 °C (Figure 21); which is exactly what we have today. In contrast if you consider the Northern Hemisphere, the continents are not on the pole but surround it. So instead of one huge ice sheet, as we have in Antarctica, there is one smaller one on Greenland, and the continents act like a fence, keeping all the sea ice in the Arctic Ocean. So the Equator–pole temperature gradient of the Northern Hemisphere is somewhere between the extremes of the Antarctic and an ice-free continent, about 50 °C. The size of the Equator–pole temperature gradient is a fundamental driver of our climate. Because the main driver of ocean and atmospheric circulation is moving heat from the Equator to the poles. So this

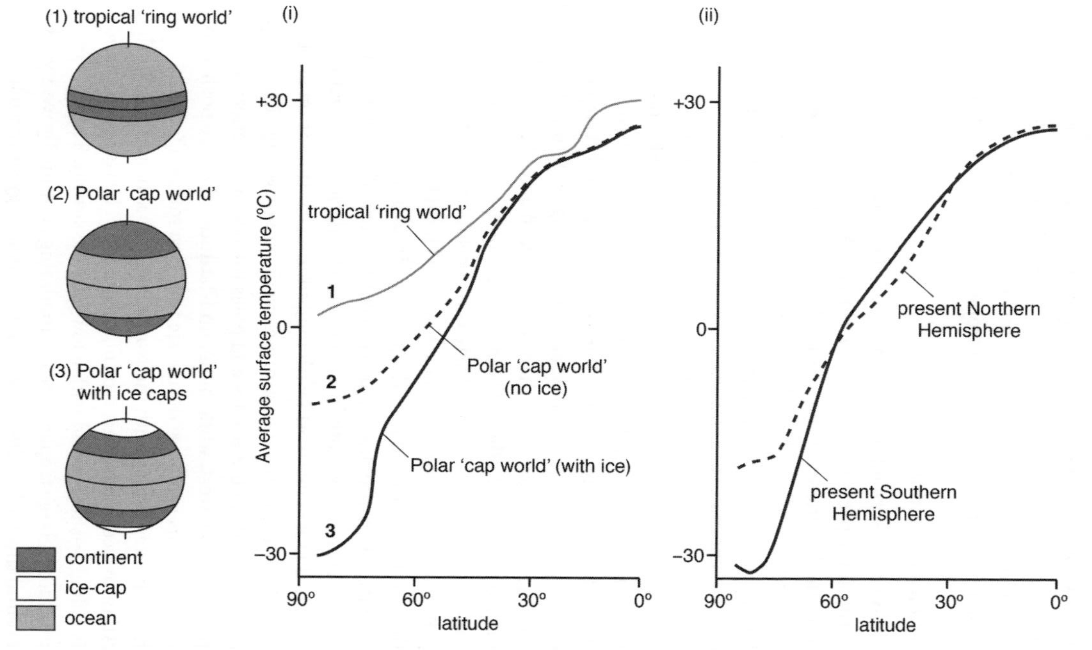

21. Latitudinal location of continent and the Equator–pole temperature gradient

temperature gradient defines what sort of climate the world will have. A cold Earth has an extreme Equator–pole temperature gradient and thus a very dynamic climate. This is why we have strong hurricanes and winter storms: the climate system is trying to pump heat away from the hot tropics towards the cold poles.

Longitude continents

Chapter 2 described the fundamentals of ocean circulation. One of the key aspects of ocean circulation is how the oceans are contained. If there are no continents in the way then oceans will just continue to circulate around and around the globe. However, when an ocean current encounters a continent it is deflected both north and south. If we look at the modern configuration of the continents (Figure 22a) then there are three main longitudinal continents: (1) the Americas, (2) Europe down to southern Africa, and (3) Northeast Asia down to Australasia. A hundred million years ago the continents are still recognizable but they are in slightly different positions (Figure 22b). The two striking features are, first, there was an ocean across the whole of the tropics through the Tethyan Sea and the Deep Central American passage. Second, there is no ocean circulating around Antarctica. These changes have huge effects on the circulation of the surface ocean and thus deep-ocean circulation and global climate. There are three main conceptual ways of understanding the effects of ocean gateways on ocean circulation. The first is a simple double-slice world with longitudinal continents on either side (Figure 23). Because ocean currents are driven by the surface winds in the tropics and poles the ocean currents are pushed to the west, while in the mid-latitudes they are pushed to the east. This produces the classic two-gyre solution in both hemispheres. Today both the North Pacific Ocean and the North Atlantic Ocean have this type of circulation. The second scenario is a double-sliced world with a low latitude seaway. This produces a large tropical ocean circulating continually westward around the world. There are then two smaller gyres in each hemisphere (Figure 23). This is the circulation seen during the

Cretaceous period, with the two gyres in each hemisphere occurring in the Pacific Ocean. The third scenario is a double sliced world with high latitude seaways. This produces strong circumpolar ocean currents in each hemisphere and a single tropical gyre in each hemisphere (Figure 23). Today the Southern Hemisphere resembles this scenario with a circumpolar current around Antarctica. The Southern Ocean thus acts like a giant ocean heat extractor and was instrumental in the huge build up of ice on Antarctica.

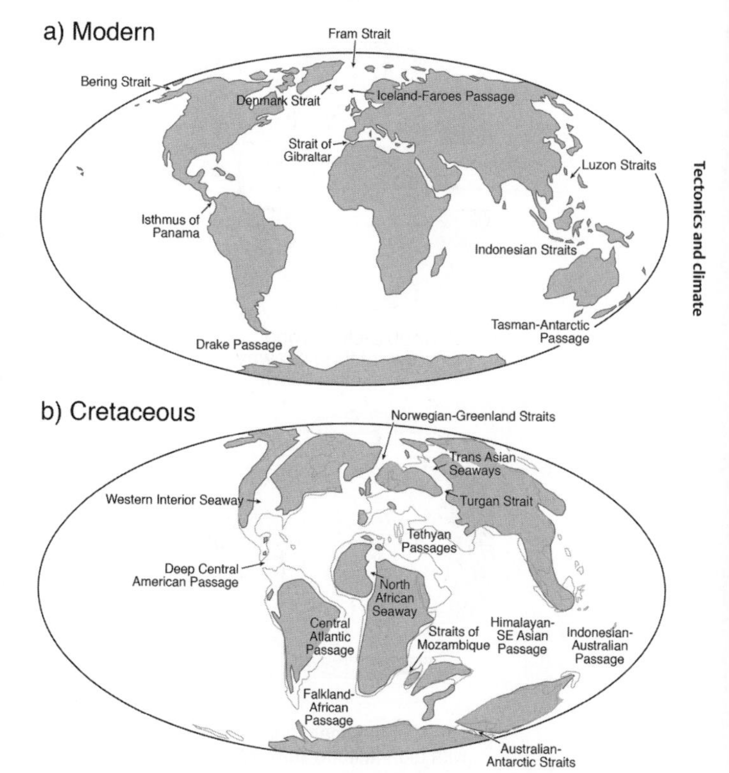

22. Ocean gateways both today and during the Cretaceous period

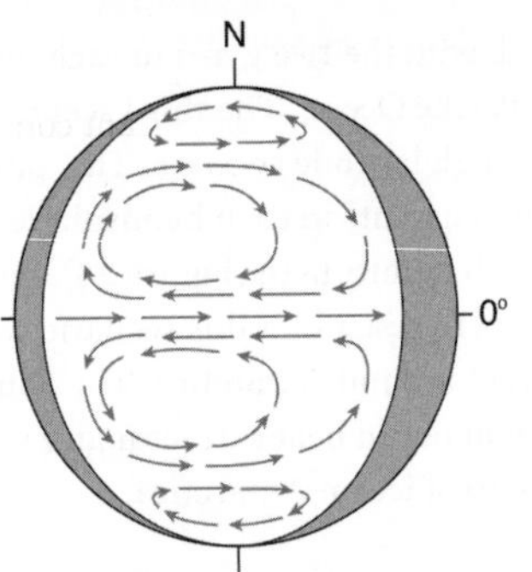

1. Double-slice world

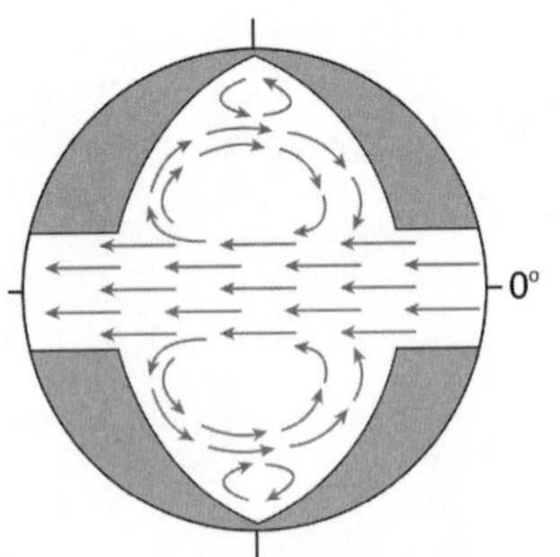

2. Double-slice world with low-latitude seaway

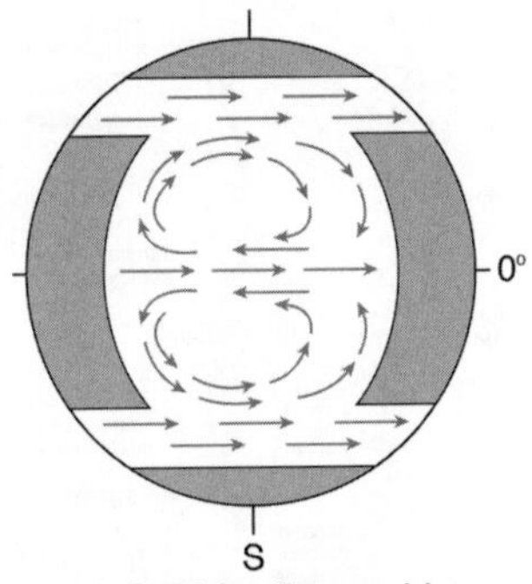

S

3. Double-slice world with high-latitude seaways

23. Longitudinal continents and ocean circulation

Deep-ocean circulation

Deep-ocean circulation is also an important consideration as it influences the circulation of the surface ocean and the distribution between the hemispheres. The presence or absence of ocean gateways has a profound effect on the deep-ocean circulation. For example, our modern day North Atlantic Deep Water (NADW), which helps to pull the Gulf Stream northwards maintaining the mild European climate may be only 4 million years old. If we run computer simulations of ocean circulation with and without the

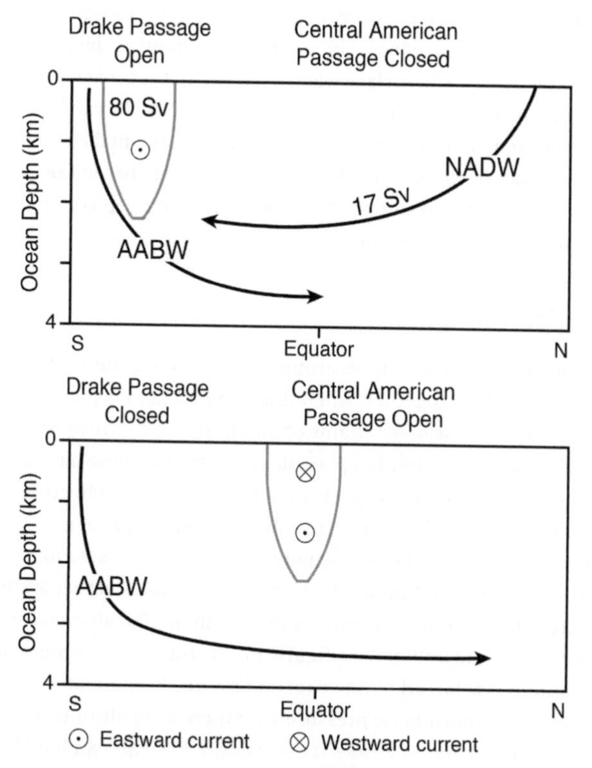

24. Ocean gateways and deep-ocean circulation

Drake Passage and the Panama Gateway, only the modern day combination produces significant NADW. Hence our modern day deep-ocean circulation is due to an open Drake Passage from about 25 million years ago and the closure of the Panama Gateway from about 4 million years later (Figure 24). It is all due to salt. Because of the greater effect of evaporation in the North Atlantic region, the North Atlantic Ocean is saltier than the Pacific Ocean. NADW forms today when the warm, salty water from the Caribbean travels across the Atlantic Ocean and cools down. The high salt load and colder temperature act together to increase the density of the water so it is able to sink north of Iceland. So when the Panama passage way is open then fresher Pacific Ocean water leaks in and reduces the overall salt content of the North Atlantic Ocean. The surface water even when it is cooled is thus not dense enough to sink and so not as much NADW can be formed compared to today. So, one of the fundamental elements of our modern climate system, the competition between the Antarctic Bottom Water and the North Atlantic Deep Water, turns out to be a very young feature.

Vertical tectonics

As the tectonic plates move around the surface of the Earth they frequently clash together, when this happens land is pushed upwards. In some cases chains of mountains are formed or when whole regions are uplifted plateaus are formed. These have a profound effect on the climate system. One of these effects is a rain shadow, which is a dry area on the leeward side of a mountain system. There is usually a corresponding area of increased precipitation on the forward side. As a weather system at ground level moves towards a mountain or plateau it is usually relatively warm and moist (Figure 25). As the air encounters the mountain it is forced to move up and over it. Because of decreasing atmospheric pressure with increasing altitude, the air has to expand and as it does it cools down. Cool air can hold less moisture than warm air so the relative humidity rapidly rises

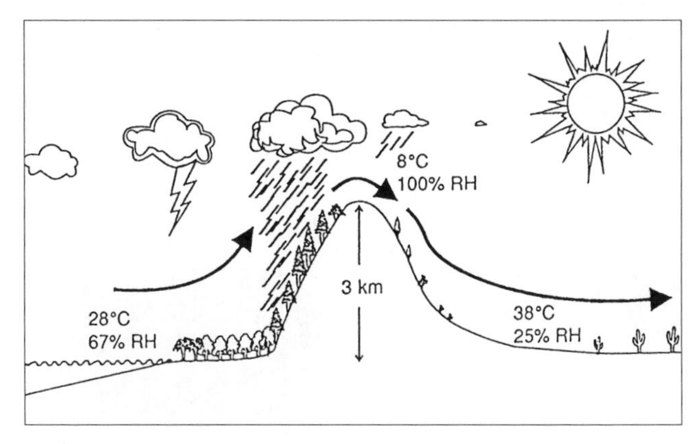

25. Mountain rain-shadow (RH = relative humidity)

until it hits 100 per cent and strong rainfall occurs. As the air descends on the other side of the mountain atmospheric pressure increases and the air temperature rises and the relative humidity drops very low as little or no moisture is left in the air. Hence on the descending side there is a rain shadow as there is no moisture left with which to form rain and this can lead to the creation of a desert. This simple process can control the wetness or dryness of whole continents. Figure 26 shows the effect of whether mountains occur on the western or eastern boundary of a continent. As we saw in Chapter 2 there are three main rainfall belts in the world, one in the tropics and one in the mid-latitudes in each hemisphere. Air in the tropics moves from east to west, while in the mid-latitudes it moves west to east. So having mountains on the western side produces more rainfall on land and produces a wetter continent overall. By coincidence at the moment we have western mountain ranges running down the west coast of North America, the Rockies, and the west coast of South America, the Andes. These mountains not only produce significant wet areas but also famous deserts like the Atacama Desert in Chile and Death Valley in the United States, which are two of the driest deserts on Earth. The contrast between wet

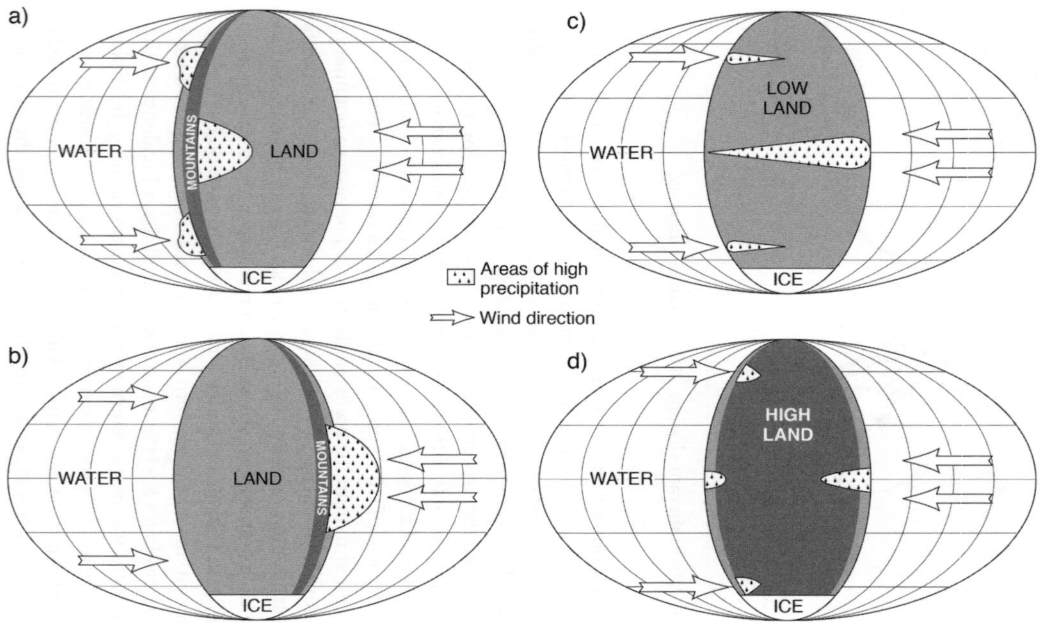

a) WATER MOUNTAINS LAND ICE

b) WATER LAND MOUNTAINS ICE

c) WATER LOW LAND ICE

d) WATER HIGH LAND ICE

Areas of high precipitation

Wind direction

26. Mountain and plateau effects on global rainfall

and dry regions is even sharper if the uplift produces a plateau. Figure 26 shows how little rainfall can make it into a plateau due to this rain shadow effect.

Atmospheric barriers

When huge mountains or plateaus are thrust high up in the sky they interfere with the circulation of the atmosphere. Not only do they force air up and over them but in many cases they deflect the weather system around them. This effect is compounded as uplift areas also warm up in summer and cool down in winter more than the surrounding lowlands. Figure 27 shows that if all the continents in the Northern Hemisphere were flat then the major circulation of the atmosphere would be nearly circular, with maybe a slight deflection due to the difference between land and the oceans. However, if you put the two modern plateaus in place, in other words, the uplifted regions of the Tibetan-Himalayan and Sierran-Coloradan plateaus then there are huge changes in circulation. Both these plateaus are massive. The Tibetan plateau is the world's highest and largest with an area of 2.5 million km^2, which is about four times the size of France. While the Colorado Plateau covers an area of 337,000 km^2 and is joined to numerous other plateaus which make up the Sierran-Coloradan uplift complex.

In Northern Hemisphere summers these two major plateaus heat up more than the surrounding areas and thus the air above them rises creating a low-pressure zone. This sucks in surrounding air creating a cyclonic circulation deflecting weather system much further north and south. In Northern Hemisphere winters these highlands are much colder than the surrounding areas creating a high-pressure system and out-blowing anti-cyclonic circulation (see Figure 27). This deflects Arctic air northwards and keeps the middle of the Asian and North American continents warmer than they would otherwise be. The atmospheric circulation becomes even more complicated when large ice sheets are present on Greenland, North America, and Europe. Because ice sheets are always cold they produce permanent high-pressure systems with out-blowing anti-cyclonic circulation,

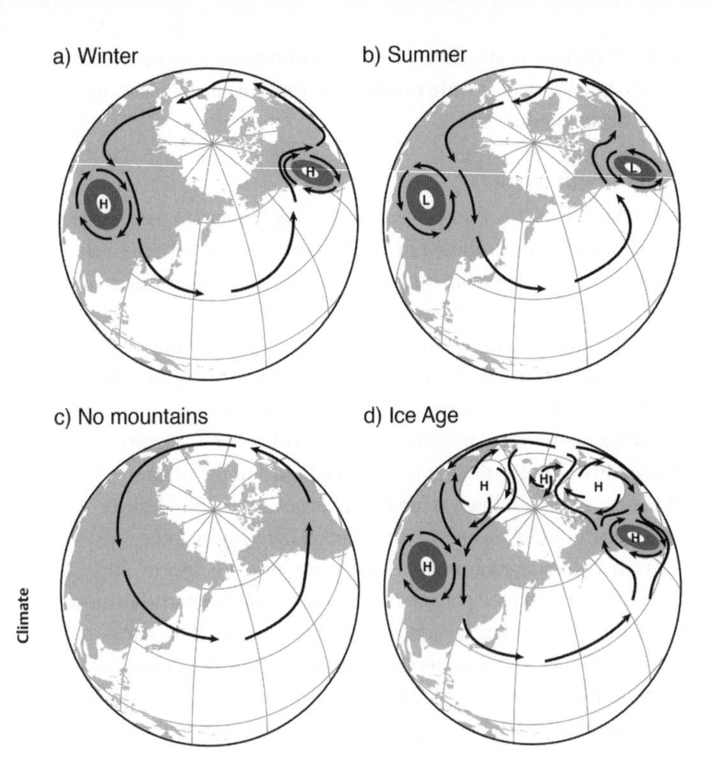

27. Plateaus and ice sheet effects on atmospheric circulation

which is discussed in Chapter 7. The summer cyclonic circulation around the Tibetan-Himalayan plateau also creates the Southeast Monsoonal system. Because part of the air that is pulled towards the Himalayas comes from the Indian Ocean it brings with it a lot of moisture. The resultant rainfall is essential for the well-being of two-fifths of the world's population.

Volcanic eruptions

Plate tectonics control the development of volcanoes, which have an important influence on climate through the introduction of gases and dust into the atmosphere. Normal sized volcanoes inject sulphur dioxide, carbon dioxide, and dust into the troposphere and can have a considerable effect on our weather. For example in 1883 Krakatoa erupted, killing 36,417 people. The eruption is considered to be the loudest sound ever heard in modern history, with reports of it being heard nearly 3,000 miles away. It was equivalent to 200 megatons of TNT, which is about 13,000 times the nuclear yield of the Little Boy bomb that devastated Hiroshima, Japan, during World War II. The sulphur dioxide and dust injected into the atmosphere increased the amount of sunlight reflected back into space and average global temperatures fell by as much as 1.2 °C in the year following the eruption. Weather patterns continued to be chaotic for years and temperatures did not return to normal until 1888.

On the 15 June 1991 Mount Pinatubo erupted sending 20,000,000 tonnes of sulphur dioxide into the atmosphere. The sulphur dioxide oxidized in the atmosphere to produce a haze of sulfuric acid droplets, which gradually spread throughout the lower stratosphere over the year following the eruption. This time modern instruments were able to measure its effects, which included a 10 per cent reduction in the normal amount of sunlight reaching the Earth's surface. This led to a decrease in Northern Hemisphere average temperatures of 0.5–0.6 °C and a global decrease in temperature of about 0.4 °C.

Both Krakatoa and Pinatubo had a short-term transient effect on climate. This is because the sulphur dioxide and dust were injected relatively low in the atmosphere and the amount of water also injected meant much of the material was washed out of the atmosphere within a few years (Figure 28). However these two eruptions are very small compared to eruptions from supervolcanoes. These are thousands of times larger than Krakatoa. They can occur when magma in the Earth rises into the crust from a hotspot but is unable to break through the crust. Pressure builds in a large and growing magma pool until the crust is unable to contain the pressure. They can also form at convergent

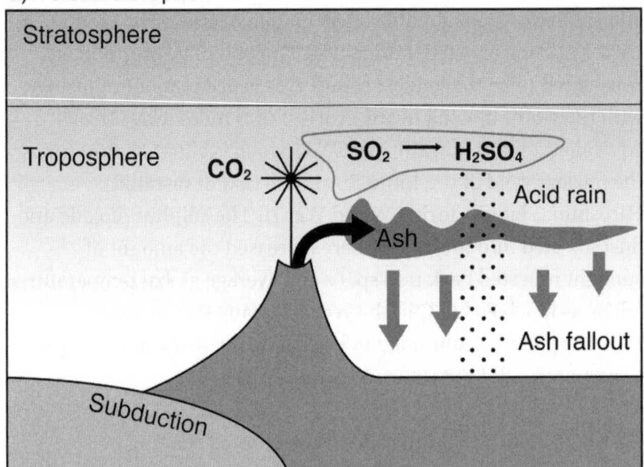

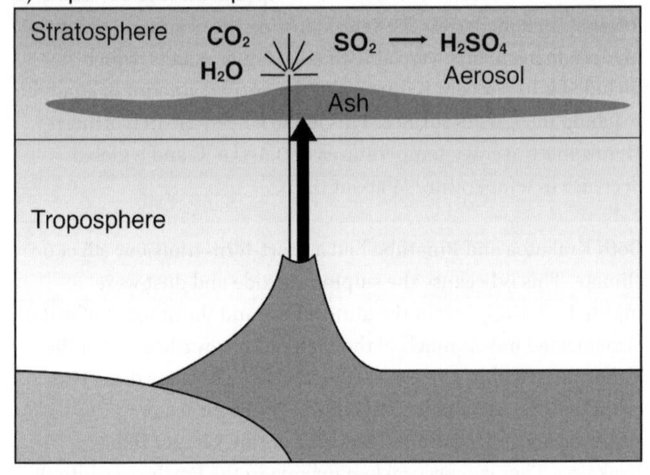

28. Volcanic eruption effects on atmosphere composition

plate boundaries, for example Toba, which last erupted about 74,000 years ago and ejected about 2,800 km^3 of material into the atmosphere. They can also form in continental hotspot locations, for example Yellowstone, which last erupted 2.1 million years ago and ejected 2,500 km^3 of material. Because of the scale of these events the sulphur dioxide and dust are injected much higher in the atmosphere and therefore the effects on the global climate can be much longer. Modelling work by the UK Meteorological Office suggested a tropical supervolcano eruption would cause a drop in global temperatures of at least 6 °C, with up to 15 °C in the tropics for at least 3 years. Then over a decade the climate would slowly came back to within 1 °C of normal. The final effects would take up to a hundred years to get rid of and would be devastating for us if it ever happened. However, in geological terms it is a very short-term event with no significant long-term effect on the climate system.

Icehouse and greenhouse worlds

Plate tectonics drives the slow shift of the continents across the globe, shifting from a supercontinent to fragmented continents and then back again. The supercontinent Rodinia formed about 1.1 billion years ago and broke up roughly 750 million years ago. One of the fragments included large parts of the continents we now find in the Southern Hemisphere. Plate tectonics brought the fragments of Rodinia back together in a different configuration about 300 million years ago, forming the best-known supercontinent, Pangaea. Pangaea subsequently broke up into the northern and southern supercontinents of Laurasia and Gondwana, about 200 million years ago. Both of these supercontinents have continued to fragment over the last 100 million years. Icehouse climates form when the continents are moving together. The sea level is low due to lack of seafloor production. The climate becomes cooler and more arid, because of the reduction in rainfall due to the strong rain shadow effect of large superplateaus. Greenhouse climates, on the other hand, are formed as the continents disperse, with sea levels high due to the high level of sea floor spreading. There are relatively high levels of carbon dioxide in the atmosphere, possibly over three times the current levels, due to production at oceanic rifting zones. This produces a warm and humid climate.

The formation and break up of these supercontinents had a huge effect on evolution. Supercontinents are extremely bad for life. First, there is a massive reduction in the amount of shelf sea areas, where we think multi-cellular life may have started. Second, the interior of continents are very dry and global climate is usually cold. A number of key mass extinctions are correlated with the formation of supercontinents. For example it is estimated that up to 96 per cent of all marine species and 70 per cent of terrestrial vertebrate species became extinct during the Permian–Triassic extinction event 250 million years ago, which is nicknamed the 'mother of all mass extinctions' (Figure 29). It is also not surprising the explosion of complex, multi-cellular organisms occurred during the Cambrian period about 550 million years ago, following the break up of the Rodinia supercontinent.

Snowball Earth

Prior to about 650 million years ago there is an idea that the surface of the Earth became entirely frozen at least once—the so-called Snowball Earth hypothesis. It is a way to explain the sedimentary deposits found in the tropics, which show glacial features that suggest there must have been a lot of ice in the tropics. Opponents of the idea suggest that the geological evidence does not suggest a global freezing. Moreover there is difficulty in getting the whole ocean to become ice- or even slush-covered. There is also the difficulty of seeing how the world, once in a snowball state, would subsequently escape the frozen condition. One answer is that this would occur through the slow build up of atmospheric carbon dioxide and methane, which would eventually reach a critical concentration, warming the atmosphere enough to start the melting process. There are a number of unanswered questions, including whether the Earth was a *full* snowball or a 'slushball' with a thin equatorial band of open water. But what is particularly

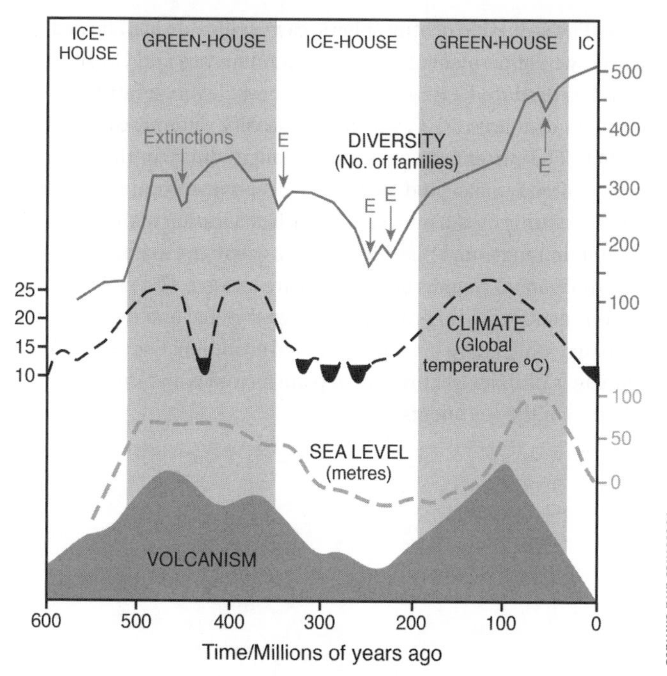

29. Long-term links between tectonics, sea level, climate, biodiversity, and extinctions

interesting is the idea that the evolution of complex life put an end to the possibilities of ever having a snowball Earth again. Professor Andy Ridgwell at Bristol University has suggested that the evolution of marine mirco-organisms that form calcite shells now buffers the oceans' carbonate system so much that the extreme variation in atmospheric carbon dioxide needed to plunge the world into or out of a snowball or slushball condition could not now occur.

Summary

Our modern climate system is a product of the slow movements of the continents across the face of the Earth. We are currently in an 'icehouse world', as we have continents on or surrounding each pole. The reduction of atmospheric carbon dioxide has allowed the growth of permanent ice sheets on Antarctica and Greenland. This has produced a very strong Equator–pole temperature gradient of at least 60 °C, which drives a very vigorous climate system. The current arrangements of longitudinal continents and ocean gateways, has produced strong deep-water formation in the North Atlantic Ocean and Antarctica. The location of modern mountain ranges and plateaus controls where the major deserts and monsoonal systems of the world are located. The movement of continents has also profoundly affected global and regional climates, which have in turn influenced evolution. Our modern climate is ultimately a product of plate tectonics and the random location of the continents.

Chapter 6
Global climate cooling

Introduction

Fifty million years ago the Earth was a very different place. The world was both warmer and wetter, with rainforest extending all the way up to northern Canada and all the way down to Patagonia. So how did we go from this lush, vibrant Earth to the ice-locked, cool planet we have today. What caused the beginning of the great ice ages? If you compare a map of the world 50 million years ago with one of the world today they seem to be the same, until you look in detail. We saw in Chapter 5 that movements of the continents around the face of the planet are very slow, but minor changes in location have had a profound effect on global climate. Over the last 50 million years these small changes have moved the Earth's climate from a being greenhouse to an icehouse world.

The last 100 million years

For the last 100 million years Antarctica has sat over the South Pole and the Americas and Asian continent have surrounded the North Pole. But only for the last 2.5 million years have we cycled in and out of the great ice ages, the so-called glacial–interglacial cycles. There must, therefore, be additional factors controlling the temperature of the Earth. In particular you need a means of cooling down the continents on or surrounding a pole. In the case of Antarctica the ice did not start building up until about 35 million years ago (Figure 30). Prior to that Antarctica was covered by lush, temperate forest: bones of dinosaurs have been found there dating from before they went extinct 65 million years ago. What changed 35 million years ago was a culmination of minor tectonic movements. Slowly South America and Australia are moving away from Antarctic. About 35 million years ago the ocean opened up between Tasmania and Antarctica. This was followed about 30 million years ago by the opening of the Drake Passage between South America and Antarctica, one of the most feared stretches of ocean. This allowed the Southern Ocean to start circulating around Antarctica. The Southern Ocean acts very much like the fluid in your refrigerator at home. It takes heat from Antarctica as it flows around the continent and then releases it into the Atlantic, Indian, and Pacific Oceans, into which it mixes. Opening up these seemingly small ocean gateways between the continents produced an ocean that can circulate around Antarctica completely, continually sucking out heat from the continent. So efficient is this process that there is now enough ice on Antarctica that if all of it melted the global sea level would rise over 65 metres—high enough to cover the head of the Statue of Liberty. This tectonic cause of the glaciation of Antarctic is also the reason that scientists are confident that global warming will not cause the eastern Antarctic ice sheet melt—if it were to melt it would cause an approximate 60 metre rise in sea level. The same cannot be said of the unstable western Antarctic ice sheet (see Chapter 8).

The ice-locked Antarctica of 30 million years ago did not, however, last long. Between 25 and 10 million years ago Antarctica ceased to be completely covered with ice. The question is why did the world start to cool all over again 10 million years ago and why did the ice start building up in the Northern Hemisphere? Palaeoclimatologists believe that relatively low levels of atmospheric carbon dioxide are essential if you are to maintain a cold planet. Computer models have shown that if you have high levels of atmospheric carbon

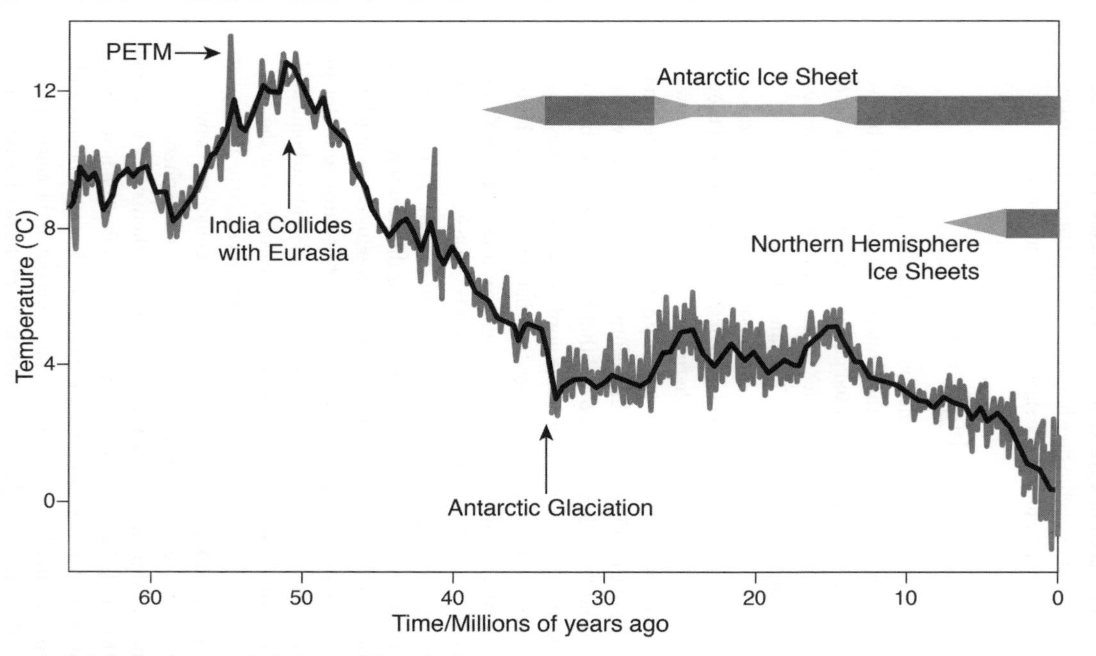

30. Global climate over the last 65 million years

dioxide you cannot get ice to grow on Antarctica even with the ocean heat extractor. So what caused the carbon dioxide to get lower and why did the ice start growing in the north?

What caused the big freeze?

In 1988 Professor Bill Ruddiman and his then graduate student Maureen Raymo while at the Lamont-Doherty Earth Observatory wrote an extremely influential paper. They suggested that global cooling and the build up of ice sheets in the Northern Hemisphere were caused by uplift of the Tibetan-Himalayan and Sierran-Coloradan regions. As we saw in Chapter 5 huge plateaus can alter the circulation of the atmosphere and they argued this cooled the Northern Hemisphere, allowing snow and ice to build up. However, what they did not realize at the time was most of the Himalayan uplift occurred much earlier between 20 and 17 million years ago and thus it was too early to have been the direct cause of the ice in the north. But Maureen Raymo then came up with a startling suggestion that this uplift may have caused a massive increase in erosion that uses up atmospheric carbon dioxide in the process. This is because when you make a mountain range you also produce a rain shadow. So, one side of the mountain has a lot more rain on it as the air is forced up and over the mountain. This is also why mountains erode much faster than gentle rolling hills. She argued that this extra rainwater and carbon dioxide from the atmosphere form a weak carbonic acid solution, which dissolves rocks. But interestingly only the weathering of silicate minerals makes a difference to atmospheric carbon dioxide levels, as weathering of carbonate rocks by carbonic acid returns carbon dioxide to the atmosphere. As much of the Himalayas is made up of silica rocks there was a lot of rock that could lock up atmospheric carbon dioxide. The new minerals dissolved in the rainwater are then washed into the oceans and used by marine plankton to make shells out of the calcium carbonate. The calcite skeletal remains of the marine biota are ultimately deposited as deep sea sediments and hence lost from the global carbon cycle for the duration of the lifecycle of the oceanic crust on which they have been deposited. It's a fast track way of getting atmospheric carbon dioxide out of the atmosphere and dumping it at the bottom of the ocean. Geological evidence for long-term changes in atmospheric carbon dioxide does support the idea that it has dropped significantly over the last 20 million years. The only problem scientists have with this theory is what stops this process. With the amount of rock in Tibet that has been eroded over the last 20 million years all the carbon dioxide in the atmosphere should have been stripped out. So there must be other natural mechanisms which help to maintain the balance of carbon dioxide in the atmosphere as the long-term concentration of carbon dioxide in the atmosphere is the result of a balance between what is removed by weathering and deposition in the deep ocean and the amount recycled by suduction zones and emitted by volcanoes.

With atmospheric carbon dioxide levels dropping between 10 and 5 million years ago the Greenland ice sheet started to build up. Interestingly Greenland started to glaciate from the south first. This is because you must have a moisture source to build ice with. So by 5 million years ago we had huge ice sheets on Antarctica and Greenland, very much like today. The great ice ages when huge ice sheets waxed and waned on North America and Northern Europe did not start until 2.5 million years ago, however there is intriguing evidence suggesting that around 6 million years ago these big ice sheets did start to grow. Rock fragments from the continent, eroded by ice and then dumped at sea by icebergs have been found in the North Atlantic Ocean, North Pacific Ocean, and Norwegian Sea at this time. This seems to have been a failed attempt to start the great ice ages and could be because of the Mediterranean Sea.

The great salt crisis

About 6 million years ago the gradual tectonic changes resulted in the closure of the Strait of Gibraltar. This led to the transient isolation of the Mediterranean Sea from the Atlantic Ocean. During this isolation the Mediterranean Sea dried out several times, creating vast evaporite (salt) deposits. Just image a huge version of the Dead Sea where a few metres of seawater cover a vast area. This event is called the Messinian Salinity Crisis and it was a global climate event because nearly 6 per cent of all dissolved salts in the world's oceans were removed. By 5.5 million years ago the Mediterranean Sea was completely isolated and was a salt desert (Figure 31). This was roughly the same time as palaeoclimate records indicate that the Northern Hemisphere was starting to glaciate. But at about 5.3 million years ago the Strait of Gibraltar reopened, causing the Terminal Messinian Flood, also known as the Zanclean Flood or Zanclean Deluge. Scientists have envisaged an immense waterfall higher than today's Angel Falls in Venezuela (979 m), and far more powerful than either the Iguazu Falls on the boundary between Argentina and Brazil or the Niagara Falls on the boundary between Canada and the USA. More recent studies of the underground structures at the Gibraltar Strait show that the flooding channel may have descended in a rather more gradual way to the dry Mediterranean. The flood could have occurred over months or a couple of years, but it meant that large quantities of dissolved salt were pumped back into the world's oceans via the Mediterranean–Atlantic gateway. This stopped the Great Ice Age in its tracks, and was entirely due to how oceans circulate. As we saw in Chapter 2 the Gulf Stream not only keeps Europe warm but also drives the deep-ocean circulation and keeps the whole planet relatively warm. Five million years ago the deep-ocean circulation was not as strong as it is today. This is because fresher Pacific Ocean water was still able to leak through the Panama ocean gateway which is discussed below. So the sudden massive increase in salt due to the Terminal Messinian Flood increased the salt in the North Atlantic Ocean ensuring a very vigorous Gulf Stream and sinking water in the Nordic Seas. With all this tropical heat being efficiently pumped northwards the slide into any further great ice ages was halted about 5 million years ago. We had to wait another 2.5 million years before the global climate was ready to try again.

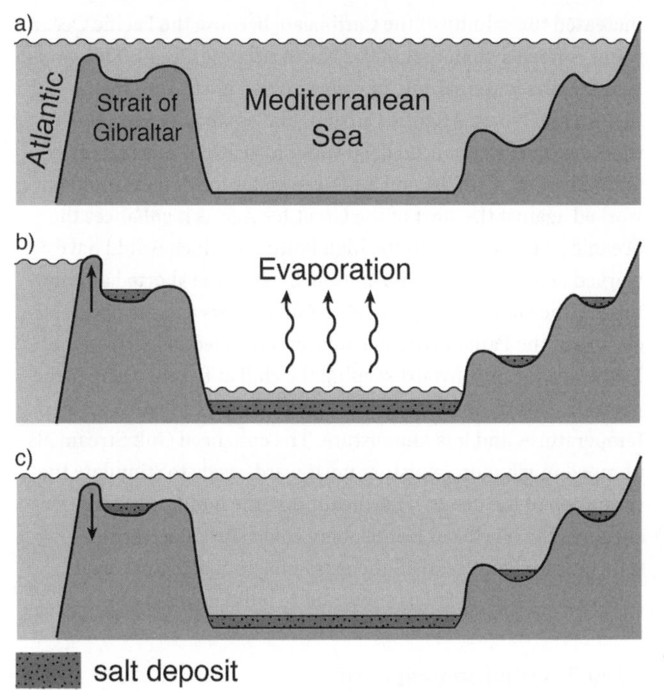

31. Mediterranean Messinian 'Salinity Crisis' and 'Terminal Flood' approximately 5 million years ago

The Panama paradox

Another important tectonic control, which geologists believe to be a trigger for causing great ice ages is the closure of the Pacific–Caribbean gateway. Professors Gerald Haug and Ralf Tiedemann now at Zurich University and the Alfred Wegener Institute used evidence from ocean sediments to suggest that the Panama ocean gateway began to close 4.5 million years ago and finally closed about 2 million years ago. The closure of the Panama gateway, however, causes a paradox, as it would have both helped and hindered the start of the Great Ice Age. First, the reduced inflow of Pacific surface water to the Caribbean would have increased the salinity of the Caribbean, because the Pacific Ocean water is fresher than that of the North Atlantic Ocean. This would have increased the salinity of water carried northward by the Gulf Stream and North Atlantic Current, and as we have seen above this would have enhanced deep water formation. The increased strength of the Gulf Stream and deep water formation would have worked against the start of the Great Ice Age as it enhances the oceanic heat transport to the high latitudes, which would have worked against ice sheet formation. So after the aborted attempt to start the Great Ice Age about 5 million years ago the progressive closure of the Panama ocean gateway continued to increase the heat transport northward keeping the chill at bay. But here is the paradox: two things are needed to build large ice sheets—cold temperatures and lots of moisture. The enhanced Gulf Stream also pumped much more moisture northward, ready to stimulate the formation of ice sheets. This meant that the building of large ice sheets in the Northern Hemisphere could start at a warmer temperature because of all the extra moisture being pumped northward ready to fall as snow and to build up ice sheets.

Why 2.5 million years ago?

Tectonic forcing alone cannot explain the amazingly fast intensification of Northern Hemisphere glaciation (Figure 32). Work of mine using ocean sediments suggests that there were three main steps in the transition to the Great Ice Ages. The evidence is based on when rock fragments that had been ripped off the continent by ice were deposited in the adjacent ocean basin by icebergs. First, sheets started growing in the Eurasian Arctic and Northeast Asia regions approximately 2.74 million years ago, with some evidence of growth of the Northeast American ice sheet; second, an ice sheet started to build up on Alaska 2.70 million years ago; and, third, the biggest ice sheet of them all, on the Northeast American continent, reached its maximum size 2.54 million years ago. So in less than 200,000 years we go from the warm, balmy conditions of the early Pliocene, which Professor

Michael Sarnthein of Kiel University called 'the golden age of climate', to the Great Ice Ages.

The timing of the start of the intensification of Northern Hemisphere glaciation must have had another cause. It has been suggested that changes in orbital forcing (changes in the way the

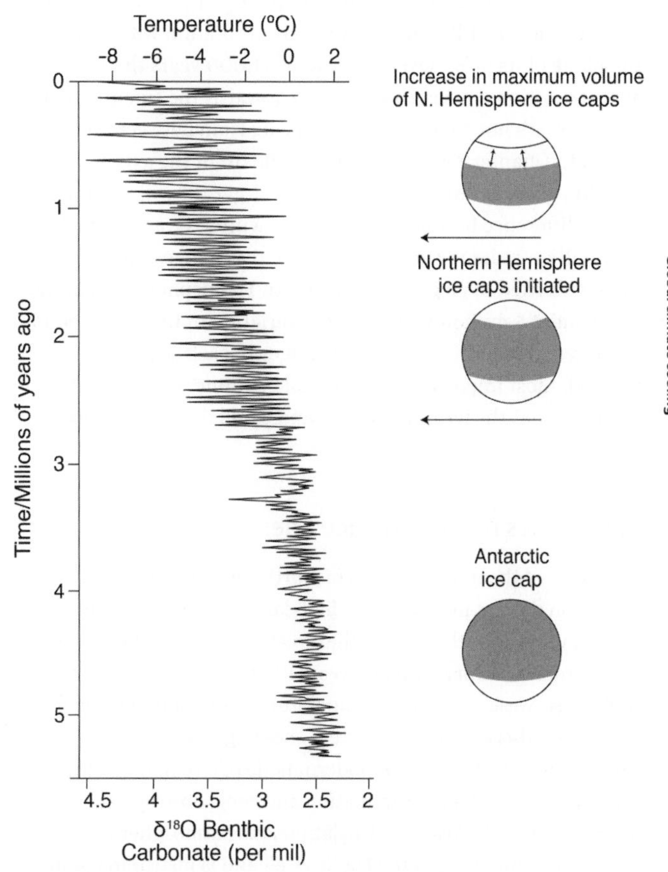

32. Global climate over the last 5 million years

Earth spins round the sun) may have been an important mechanism contributing to the global cooling. The details of the Earth's numerous wobbles and how they caused the waxing and waning of individual ice ages are discussed in the next chapter. But though these individual wobbles are on the scale of tens of thousands of years, there are much longer variations as well. For example one of the most important is obliquity or tilt, which is the wobble of the Earth's axis of rotation up and down—or, put another way, the tilt of the Earth's axis of rotation with respect to the plane of its orbit. Over a period of 41,000 years the Earth's axis of rotation will lean a little more towards the sun and then a little less. It's not a large change, varying from between 21.8° and 24.4°. In Chapter 1 we explored how the tilt of the axis of rotation gives us our seasons. Hence a larger tilt will result in a larger difference between summer and winter. Over a period of 1.25 million years the amplitude of the tilt changes. Both times the Earth tried to glaciate the Northern Hemisphere, at 5 million years and 2.5 million years ago, the variation of tilt increased to its largest value. This made the changes in each season very marked, most important were the cold summers in the north ensuring that the ice did not melt and could develop into ice sheets.

The tropics react to the ice ages

The onset of the intensification of Northern Hemisphere glaciation did not just affect the high latitudes. It seems half a million years after the start of the Great Ice Ages things changed in the tropics. Before 2 million years it seems there was a very slight east–west sea-surface temperature gradient in the Pacific Ocean, but this gradient later grew, showing a switch in the tropics and sub-tropics to a modern mode of circulation with relatively strong Walker Circulation and cool sub-tropical temperatures. The Walker Circulation is the atmospheric east–west component of the Hadley Cell and is instrumental in controlling rainfall in the tropics. The Walker Circulation is also a key element in the El Niño–Southern Oscillation (ENSO; see Chapter 3). So before 2 million years ago ENSO may not have existed in its modern form because there was a relatively weak Walker Circulation. The development of the Walker Circulation also seems to be linked to early human evolution. The strengthening of the east–west circulation seems to have produced deep, fresh but ephemeral lakes in the East African Rift. It has recently been postulated that this distinctive climate pulse with rapid appearance and disappearance of lakes at about 2 million years may be linked with the evolution of *Homo erectus* in Africa with an over 80 per cent increase in brain size and our ancestors migrating out of Africa for the very first time.

Mid-Pleistocene Transition

Mid-Pleistocene Transition (MPT) is the marked prolongation and intensification of glacial–interglacial climate cycles that occurred sometime after 800,000 years ago (Figure 32). Prior to the MPT, the glacial–interglacial cycles seem to occur every 41,000 years, which corresponds to the slow changes in the tilt of the Earth. After about 800,000 years ago the glacial–interglacial cycles seem to be much longer, averaging over 100,000 years. The shape of these cycles also changes. Before the MPT the transition between glacial and interglacial periods is smooth and the world seemed to spend about equal time in each climate. After the MPT the cycles became saw-toothed with ice building up over 80,000 years to produce deep, intense ice ages and then rapid deglaciation, with the loss of all that ice within 4,000 years. The climate then stayed in an interglacial period resembling our current climate for about 10,000 years before descending back into an ice age. One suggestion for this saw-toothed pattern is that the much larger ice sheets are very unstable and therefore with a slight change in climate they collapsed rapidly and the whole climate system rebounded back into an interglacial period. In the next chapter we will examine these recent glacial–interglacial cycles in more detail.

Chapter 7
Great ice ages

Introduction

In 1658 Archbishop Ussher of Armagh looked at the features of the landscape around him and attributed them to Noah's Flood. Using the Bible he diligently dated the flood and thus the landscape to 4004 BC. It was not until 1787 that Horace-Bénédict de Saussure, a Genevan aristocrat, physicist, and Alpine traveller, recognized that Alpine erratic boulders had been moved hundreds of miles down the slopes of the Jura Range and reasoned that the mountain glaciers must have extended much further into the past. This discovery had to wait until 1837 for Louis Agassiz, the Swiss geologist, to put forward his 'ice age' or 'glacial' theory based on the evidence of erratic boulders and end moraines. Terminal moraines are hills formed by eroded sediment being pushed in front of an advancing ice sheet like a bulldozer. When the ice sheet reaches its maximum extent the sediment is deposited as a line of hills which trace the front edge of the ice sheet. In 1909 Albrecht Penck and Eduard Brückner, German and Austrian geographers, produced a 3-volume work entitled *Die Alpen im Eiszeitalter* (*The Alps in the Ice Age*). They concluded that there were four major 'ice age' or glacial periods, the Gunz, Mindel, Riss, and Wurn. Terrestrial or land based evidence has the disadvantage of being discontinuous and the evidence may be destroyed by subsequent ice sheet advances. Hence it was not until the 1960s when long, continuous sediment cores were recovered from the bottom of the oceans that it was realized how many ice ages there had been. We now have the ability to drill in the ocean to a depth of four miles and still be able to recover over half a mile of sediment below the sea floor. From studying these marine sediments scientists have documented 50 ice ages that occurred in the last 2.5 million years.

Waxing and waning of the great ice ages

We now know that glacial–interglacial cycles are the fundamental characteristic of the Quaternary Period, the last 2.5 million years. The waxing and waning of the huge continental ice sheet is initiated by the changes in the Earth's orbit around the sun. The Earth over long periods of time wobbles on its axis and thus changes the amount of sunlight or solar energy received by different parts of the Earth. These small changes are enough to push or force climate change. However, these waxings and wanings are not caused by Earth's orbital wobbles, but rather by the Earth's climate reaction, which translates relatively small changes in regional solar energy into major climatic variability. For example, the position of the Earth is very similar today to what it was 21,000 years ago during the last ice age. So it is not the exact orbital position that controls the climate but rather the changes in the orbital positions. There are three main orbital parameters or wobbles, called eccentricity, obliquity or tilt, and precession (see Box 4), and as you can see from the box each has a unique cycle and effect on climate. More exciting though is when we combine them all together and see how they push the climate either into or out of a great ice age.

Clockwork climate?

Combining the effects of all three orbital parameters you can calculate the solar energy received for any latitude back through time. Milutin Milankovitch, a brilliant Serbian mathematician and climatologist, in 1949 suggested that summer insolation at

Box 4 Orbital forcing

There are three main orbital parameters or wobbles: eccentricity, obliquity (tilt), and precession (Figure 33), which have a significant effect on the long-term climate of the Earth.

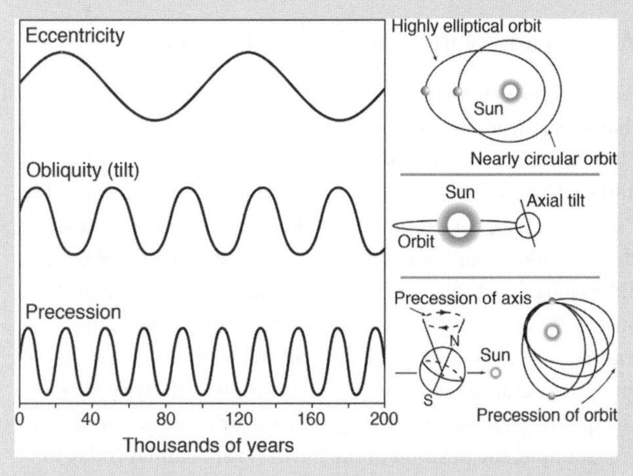

33. Orbital variables

Eccentricity is the shape of the Earth's orbit around the sun, which varies from a circle (precession) to an ellipse. These variations occur over a period of about 96,000 years with an additional long cycle of about 400,000 years. Described another way, the long axis of the ellipse varies in length over time. In recent times, the Earth is closest to the sun on 3 January at about 146 million km; this position is known as perihelion. On 4 July the Earth is at its greatest distance from the sun, at about 156 million km; this position is known as the aphelion. Changes in eccentricity cause only very minor variations in total insolation, but can have significant seasonal effects when combined with precession.

If the orbit of the Earth were perfectly circular there would be no seasonal variation in solar insolation. Today, the average amount of radiation received by the Earth at perihelion is ~351 Wm^2 and is 329 Wm^2 at aphelion. This represents a difference of ~6 per cent, but at times of maximum eccentricity (ellipse length) over the last 5 million years the difference could have been as great as 30 per cent. Milutin Milankovitch in 1949 suggested that northern ice sheets are more likely to form when the sun is more distant in summer, so that each year some of the previous winter's snow survives. The other effect of eccentricity is to modulate the effects of precession. However, it is essential to note that eccentricity is by far the weakest of all three orbital parameters.

Obliquity or the tilt of the Earth's axis of rotation with respect to the plane of its orbit varies between 21.8° and 24.4° over a period of 41,000 years. It is the tilt of the axis of rotation that causes the seasons, as described in Chapter 1. The larger the obliquity, the greater the difference between the insolation received in summer and winter. Milutin Milankovitch suggested that the colder the Northern Hemisphere summers, the more likely it is that snowfall will accumulate, resulting in the gradual build up of glaciers and ice sheets.

Precession has components that relate to the elliptical orbit of the Earth (eccentricity) and its axis of rotation. The Earth's rotational axis precesses every 27,000 years. This is similar to the gyrations of the rotational axis of a toy spinning top. Precession causes a change in the Earth–sun distance for any particular date, for example, the beginning of the Northern Hemisphere summer. It is the combination of the different orbital parameters that lead to the two different precessional periodicities of 23,000 years and 19,000 years. Combining the precession of the axis of rotation plus the precessional changes in orbit produces a period of 23,000 years. However, the combination of eccentricity (96,000 years) and precession of the axis of rotation also results in a period of 19,000 years. These two periodicities combine so that perihelion coincides with the summer season in each hemisphere on average every 21.7 thousand years. Precession has the most significant impact in the tropics (in contrast to the impact of obliquity at the Equator, which is zero). So although obliquity clearly influences high latitude climate change, which may ultimately influence the tropics, the direct effects of insolation in the tropics are due to eccentricity-modulated precession alone.

Combining the effects of eccentricity, obliquity, and precession provides the means for calculating insolation for any latitude through time. Figure 34 shows the calculated insolation for 65°N compared with the changing size of the ice sheets represented by global sea level change for the last 600,000 years.

65°N which is just south of the Arctic Circle was critical in controlling glacial–interglacial cycles (Figure 34). He argued that if the summer insolation was reduced enough then ice could survive through the summer, start to build up, and eventually produce an ice sheet. Orbital forcing does have a large influence on this summer insolation; the maximum change in solar radiation in the last half million years is equivalent to reducing the amount of summer radiation received today at 65°N to that received now over 550 km to the north at 77°N. In simplistic terms, this brings the current ice limit in mid-Norway down to the latitude of mid-Scotland. These lows in 65°N insolation are caused by eccentricity elongating the summer Earth–sun distance, obliquity being shallow and precession placing the summer season at the longest Earth–sun distance produced by eccentricity. The reason why it is at 65°N and not at 65°S, which controls climate, is very simple. Any ice that builds up in the Northern Hemisphere has many continents to grow upon. In contrast in the Southern Hemisphere the ice growth is limited by the Southern Ocean, as

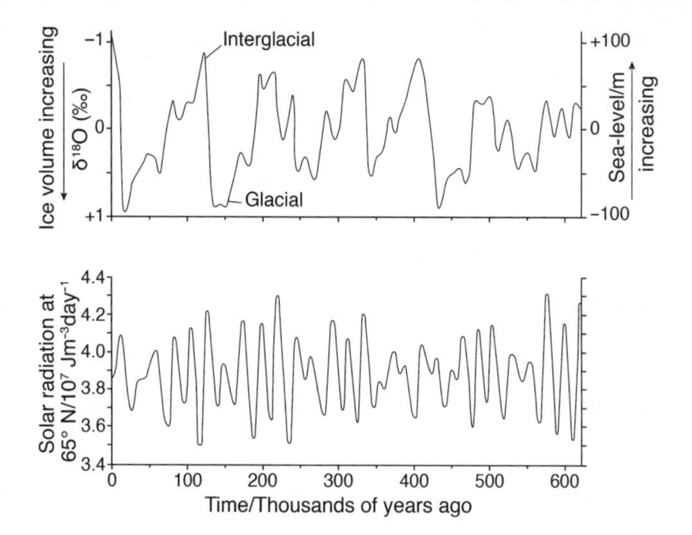

34. Comparison of Northern Hemisphere solar radiation and global sea level

any extra ice produced on Antarctica falls into the ocean and is swept away to warmer seas (Figure 35). So the conventional view of glaciation is that low summer solar energy in the temperate Northern Hemisphere allows ice to survive summer and thus ice sheets start to build up on the northern continents. But this apparently simple clockwork view of the world is really much more complicated, as the effects of orbital changes on the seasons is very small and it is feedback in the climate system that amplifies these changes.

What causes the glacial–interglacial cycles

Orbital forcing in itself is insufficient to drive the observed glacial–interglacial variability in climate. Instead, the Earth system amplifies and transforms the changes in solar energy received at the Earth's surface through various feedback mechanisms. For example, let us start with building an ice age.

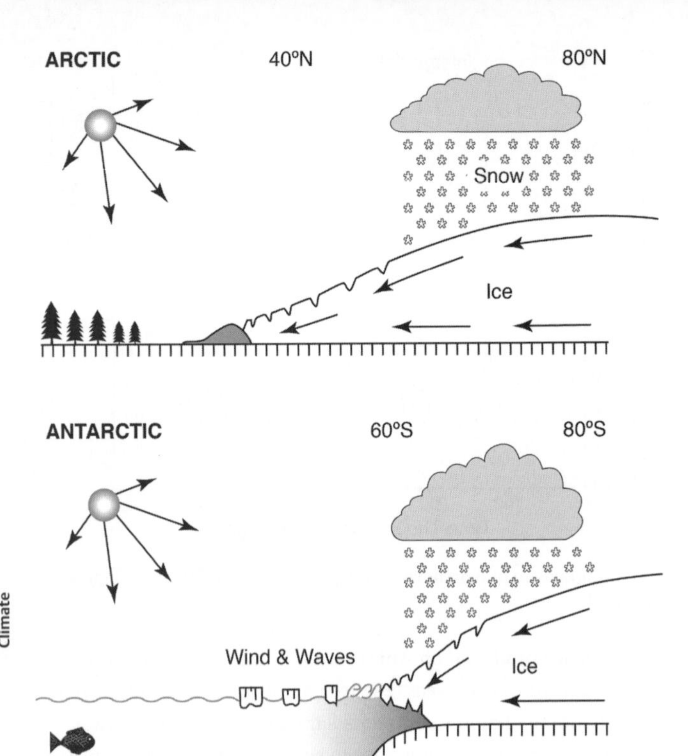

35. Ice sheet expansion in the Arctic and Antarctic

The first thing that needs to happen is a slight reduction in summer temperatures. As snow and ice accumulate due to this change in summer temperature, the albedo, the reflection of sunlight back into space, increases. The process of reflecting more sunlight back into space suppresses local temperatures, which promotes the accumulation of more snow and ice, which further increases the albedo of the region, producing the so-called 'ice–albedo' feedback. So once you have a small ice sheet it changes the environment around it to make more snow and ice, and will get bigger and bigger.

Another feedback is triggered when the ice sheets, particularly the Laurentide ice sheet on North America, become big enough to deflect the atmospheric planetary waves (see Chapter 5, Figure 27). This changes the storm path across the North Atlantic Ocean and prevents the Gulf Stream and North Atlantic Drift from penetrating as far north as it does today. This surface ocean change combined with the general increase in melt-water in the Nordic Seas and Atlantic Ocean due to the presence of a large continental ice sheet ultimately leads to a reduction in the production of deep water. Deep-water production in the Greenland and Labrador Seas is the heartbeat of our modern climate. By reducing the formation of deep water it reduces the amount of warm water pulled northwards, all of which leads to increased cooling in the Northern Hemisphere and expansion of the ice sheets.

There is currently a debate amongst palaeoclimatologists about the role of the physical climate feedbacks described above and the role of greenhouse gases in the atmosphere. Air bubbles trapped in polar ice have shown us that carbon dioxide dropped by a third and methane by a half during each glacial period. These changes would have compounded the cooling that occurred during each glacial period, helping to build more ice. So the debate continues: do changes in the Earth's orbit affect the production of greenhouse gases, cooling down the Earth to make the Northern Hemisphere continents susceptible to the build up of large ice sheets? Or do changes in the Earth's orbit start the build up of large ice sheets in the Northern Hemisphere that then change global climate and reduce the production of greenhouse gases, thus prolonging and deepening the ice age? I am sorry to say the jury is still out on these questions. However, what we do know is that greenhouse gases played a critical role in glacial—interglacial cycles. We also know that changes in greenhouse gases concentration always come before changes in global temperatures.

An important question is why these feedbacks don't end up running away and freezing the whole Earth. The answer is that they are prevented from having a runaway effect by a process called 'moisture limitation'. To build an ice sheet you need it to be cold and wet. However, as the warm surface water is forced further and further south, supply of the moisture that is required to build ice sheets decreases. So by changing the atmospheric and ocean circulation, the ice sheets end up starving themselves of moisture.

In the last million years it took up to 80,000 years to build up ice sheets to reach the maximum extent of ice. So the last time this occurred was about 21,000 years ago. However, getting rid of the ice is much quicker. This process known as 'deglaciation' usually takes a maximum of only 4,000 years. This deglaciation is triggered by an increase in solar energy received during the summer at about 65°N. This encourages the Northern Hemisphere ice sheets to melt slightly. The rise of atmospheric carbon dioxide and methane promotes warming globally and encourages the melting of the large continental ice sheets. But these processes have to work against the ice sheets' albedo effect, which produces a microclimate that works to keep them intact. What causes the rapid removal of ice is the rise in sea level due to the melting ice sheet, where large ice sheets adjacent to the oceans are undercut by a rising sea level. The coldest sea water can become is about −1.8°C while the base of the ice sheet is usually colder than −30°C. The effect of the water against the ice is similar to putting hot water under a tub of ice cream. Undercutting of the ice sheet leads to more melting and ice caving into the ocean. This in turns increases sea level, which causes even more undercutting. This sea level feedback process can occur extremely rapidly. Once the ice sheets are in full retreat then the other feedback mechanisms discussed above are thrown into reverse.

Anatomy of the last ice age

If we focus on the last ice age, only 21,000 years ago, we can see what a considerable difference glaciation made to the climate of Earth. In North America there was nearly continuous ice spreading across the continent from the Pacific to the Atlantic Ocean. It was made up of two separate ice sheets, the Laurentide ice sheet in the east, centred on Hudson Bay, and the Cordilleran ice sheet in the west, in the coastal ranges and the Rockies. The Laurentide covered over 13 million km^2 of land and reached over 3,300 m in thickness at its deepest over the Hudson Bay. Its maximum reach extended from New York, to Cincinnati, to St Louis, and to Calgary. In Europe there were two major ice sheets, the Fenno-Scandinavian and the British ice sheets, with a minor one over the European Alps. The British ice sheet during many glacial periods merged with the Scandinavian ice sheet. On average it covered about 340,000 km^2 per glacial period. During the last glacial period the ice sheet reached halfway down the British Isles to the edge of Norfolk. The Fenno-Scandinavian ice sheet was much larger than the British ice sheet, covering an area of 6.6 million km^2, and extended all the way from Norway to the Ural Mountains in Russia. We must also not forget the Southern Hemisphere as there were significant ice sheets in Patagonia, South Africa, southern Australia, and New Zealand. In addition the Antarctic ice sheet expanded by about 10 per cent and seasonal sea ice extended an additional 500 miles away from the continent. It is difficult to imagine the huge amount of water that was locked up in these ice sheets. One way of understanding is to consider the oceans. The oceans cover over 70 per cent of our planet and so much water was sucked out of them and locked up in the ice sheets that the sea level dropped by over 120 m. This is approximately the equivalent of the height of the London Eye. If all the ice on Antarctica and Greenland today melted it would raise the sea level by 70 m. During a glacial period global temperatures were 5–6 °C lower than today, but this temperature drop was not evenly distributed, with the high latitudes cooling by as much as 12 °C, while even the tropics cooled by between 2 °C and 5 °C. The ice ages were also very dry periods with large amounts of dust in the atmosphere. For example, in Northern China, eastern USA, central and eastern Europe, central Asia, and Patagonia there are deposits of hundreds of metres of dust called 'loess deposits', which built up during the ice age.

Ice shapes the land

During the glacial periods the presence of these huge ice sheets profoundly affected the local climate, vegetation, and landscape. The great boreal forests of the high latitudes were devastated as the land they once occupied was overrun by expanding ice sheets. The reduction of atmospheric moisture greatly reduced rainfall and the great wetlands and tropical rainforest of the world shrank. The huge continental ice sheets also had a profound effect on our landscape. There are very few places in the temperate latitudes that have not been affected by the ice ages. If you travel through northern Europe and North America you will see the dramatic effects of ice ages on the landscape. These effects have made great backdrops for movies, for example the *Lord of the Rings* trilogy was filmed in New Zealand and the mountainous, wild, extreme landscape you see is the result of an ice sheet grinding away over the islands for thousands of years. So, next time you see those movies, think 'ice sheets'. The ice sheets have left us with a legacy of U-shaped valleys, fiords, moraines, and egg shaped hills called 'drumlins'. Even the current position of the River Thames is due to the ice. Previously the River Thames ran through St Albans to the north of London and met the North Sea in Essex. The last but one ice age was so intense in Europe that the ice sheet made it down as far as north London. This re-routed the River Thames to its current path. So the geography of London was primarily controlled by the ice age. In the USA the paths of many major rivers were altered both by the location of an ice sheet and also by the huge amount of meltwater which burst from them as they later melted 12,000 years ago. The pathways of the Laurentide and Mississippi rivers are relics of these great floods at the end of the last glacial period.

The geography of the Earth was also altered as the lowering of global sea level by 120 m meant that the continents changed shape. Islands such as Britain became part of the mainland, meaning that during the last ice age it would have been possible to walk across the English Channel to France. The only thing that would stop you would be the huge new river running down the centre of what is now the English Channel taking water from the Thames, the Rhine, and the Seine out to the Atlantic Ocean. All around the world land bridges were formed by the lowering sea level allowing new species to invade new areas. Around the world, islands such as Sri Lanka, Japan, Britain, Sicily, Papua New Guinea, and the Falklands became part of the adjacent mainland. For example, the chain of islands across the Bering Sea, which separates Northeast Asia from Alaska became joined. So during the end of last glacial period, as the climate started to warm up, humans were able to cross from Asia into North America for the very first time, colonizing the New World.

The case of the missing grass in Amazonia

Ice ages clearly affect the whole of the global climate system, however there is controversy concerning what effects they had on the tropics. Half the surface of the planet lies between the Tropics of Cancer and Capricorn and includes all the tropical rainforest of the world. Of these the most important region in terms of size and species' diversity is the Amazon. The Amazon Basin is the largest in the world covering an area of 7 million km^2 and it discharges approximately 20 per cent of all freshwater carried to the oceans. The majority of it is covered by extremely diverse rainforest.
In 1969 Haffer put forward a wonderful theory linking the ice age

to why the Amazon is so diverse. During each glacial period, he suggested, lower temperatures and precipitation in the tropics allowed savannah to replace the majority of the tropical rainforest. However some of the tropical rainforest would have survived in small 'refugia', isolated islands of rainforest surrounded by grassland. These isolated patches of rainforest would have become hotbeds of evolution, producing many new species. At the end of each glacial period the patchwork of rainforest merged back together with higher levels of species diversity and endemism than previously. However, by the late 1990s this theory came under attack, as more and more scientists failed to find the huge increase in savannah. We now know from pollen records and computer models that in the Amazon the combination of dry and cold conditions meant that savannah did encroach a little bit at the edges, reducing the area of the Amazon rainforest to 80 per cent of today's coverage. But it is a testament to the resilience and importance of tropical rainforest in the global ecosystem that the Amazon survived and even flourished during an ice age. One of the reasons why the rainforest survived the glacial period was because the cold conditions actually helped to reduce the problem of less rainfall: the cold temperatures reduced the amount of evaporation from the trees and thus the loss of moisture which is essential for the rainforest. However, there were major changes in the species composition of the Amazon rainforest during glacial periods. For example, we know from pollen records that many of the tree species now found in the Andes were in the heart of the Amazon forest. This is because the species which are more cold-adapted are pushed up to higher 'colder' altitudes during warm interglacial periods. This is important as it means we cannot see the current Amazon rainforest as the normal condition, because for the last million years the Earth's climate has spent about 80 per cent of the time in glacial conditions. So the Amazon forest reconstructed for the last glacial period with a diverse mix of Andean and lowland tropic tree species and evergreen and partial evergreens is the norm. The lack of grassland in the Amazon during glacial periods also means we have to look for other evolution mechanisms for the huge diversity of the Amazon rainforest, and it may be that the ice ages were not the cause.

Unstable ice ages

In many ways glacial periods really should be called 'climate rollercoasters', because ice sheets are naturally unstable and during glacial periods the climate veers violently from one state to another as the ice sheets dramatically collapse and then reform. Most of the variations occur on a millennial time-scale, but the start of these extreme events can occur in as little as 3 years. The most impressive of these events are the Heinrich events. These events were named by Professor Wally Broecker, a palaeoceanographer at Lamont-Doherty Earth Observatory, after Hartmut Heinrich who wrote a paper describing them in 1988. Heinrich events are massive collapses of the North America Laurentide ice sheet that result in millions of tonnes of ice being poured into the North Atlantic Ocean. Wally Broecker described them as armadas of icebergs floating from North America across the Atlantic Ocean to Europe. Huge gouges have been found on the north French coast where these huge icebergs ran aground. These Heinrich events occurred against the general background of an unstable glacial climate, and represent the brief expression of the most extreme glacial conditions around the North Atlantic region. The Heinrich events are evident in the Greenland ice core records as a further 2–3 °C drop in temperature from the already cold glacial climate. The Heinrich events have been found to have had a global impact, with evidence of major climate changes described from as far afield as South America, the North Pacific, the Santa Barbara Basin, the Arabian Sea, China, the South China Sea, and the Sea of Japan. During these events around the North Atlantic region much colder conditions are found both in North America and Europe. In the North Atlantic Ocean the huge number of melting icebergs added so much cold fresh water that the sea-surface temperatures and salinity were reduced to the extent that surface water could not sink. This stopped all deep water formation in the North Atlantic Ocean switching off the global ocean conveyor belt.

Heinrich events are easy to spot in marine sediment cores from the middle of the Atlantic Ocean. This is because the icebergs bring huge amounts of rock with them into the ocean and as they melt they leave a trail of rock fragments scattered over the floor of the ocean. By recognizing these events in ocean sediments and dating the fossil in the sediment, it seems the Heinrich events occurred on average every 7,000 years during the last ice age. Also below these rock fragments we have found little burrows of marine worms. Usually these burrows cannot be seen as the sediment is mixed up by other animals coming to feed on it. For these fossil tubes and burrows to be preserved then a rain of rock fragments from the melting icebergs must have occurred within 3 years and been rapid enough to prevent other animals getting to the sediment. This evidence suggests that the collapse of the North American ice sheet was extremely rapid with icebergs flooding the Atlantic Ocean in less than 3 years. So, during an ice age conditions varied from cold conditions with massive ice sheets to extreme cold conditions brought on by the partial collapse of the North American ice sheet.

We now know that between the massive Heinrich events there are smaller events occurring about every 1,500 years, which are referred to as Dansgaard-Oescheger events or cycles. One suggestion is that Heinrich events are in fact just super Dansgaard-Oescheger events. The big difference between Heinrich events and Dansgaard-Oescheger events is that Heinrich events are only found during ice ages, while Dansgaard-Oescheger events have been found both in interglacial as well as in glacial periods.

What caused the Heinrich events?

The fascination with the Heinrich events is due to their occurrence on a time-scale that we can appreciate and the massive and profound effect they had on the climate of an ice age. So there are lots of competing theories for what caused them. A glaciologist called Doug MacAyeal suggested that the Heinrich event iceberg surges were caused by internal instabilities of the Laurentide ice sheet. This ice sheet rested on a bed of soft, loose sediment; when it is frozen it does not deform, behaving like concrete, and so would have been able to support the weight of the growing ice sheet. As the ice sheet expanded the geothermal heat from within the Earth's crust together with heat released from friction of ice moving over ice was trapped by the insulating effect of the overlying ice. This 'duvet' effect allowed the temperature of the sediment to increase until a critical point when it thawed. When this occurred the sediment became soft, and thus lubricated the base of the ice sheet causing a massive outflow of ice through the Hudson Strait into the North Atlantic. This, in turn, would lead to sudden loss of ice mass, which would reduce the insulating effect and lead to re-freezing of the basal ice and sediment bed, at which point the ice would revert to a slower build up and outward movement. Doug MacAyeal called this a binge–purge model and suggested all ice sheets have their own times of instability, thus the Fenno-Scandinavian, Greenland, and Icelandic ice sheets would have surges with different periodicities.

Another exciting theory is the 'bipolar climate see-saw' idea—another wonderful term created by Professor Wally Broecker. This theory is based on new evidence from the ice cores in Greenland and Antarctic, which show that during Heinrich events the climate of the Northern and Southern Hemispheres are out-of-phase. So when the climate of the Northern Hemisphere is cooling down it is warming up in Antarctica. It has been suggested that this so-called bipolar climate see-saw can be explained by alternating ice sheet collapse and resultant melt-water events in the North and South Atlantic Ocean. Each melt-water event would change the relative amount of deep-water formation in the two hemispheres and the resulting direction of the inter-hemispheric heat piracy. At the moment the Northern Hemisphere steals heat from the Southern Hemisphere to maintain the Gulf Stream and the relatively warm deep-water formation in the Nordic Seas. The heat is slowly returned by the flow of deep water from the North to the South Atlantic Ocean. So the bipolar climate see-saw model suggests if the ice sheets around the North Atlantic collapsed sending huge amounts of icebergs into the ocean they will melt. This melt water would make the ocean too fresh so none of the water could sink. This stops the formation of North Atlantic Deep Water and the Northern Hemisphere stops stealing heat from the Southern Hemisphere. This results in the Southern Hemisphere slowly warming up. Over maybe 1,000 years this heat build up is enough to collapse the edges of the Antarctic, which then shuts off the deep-water formation around Antarctica and the whole system is reversed. The nice thing about this theory is it can work in an interglacial period as well, and as we saw above the Dansgaard-Oescheger cycles of about 1,500 years occur during both glacial and interglacial periods.

Holocene

We are currently in an interglacial period called the Holocene, having escaped the last ice age about 10,000 years ago. The climate has not been constant during our interglacial period, and the early Holocene period may have been warmer and wetter than the 20th century. Throughout the Holocene there have been millennial-scale climate events, Dansgaard-Oescheger cycles, which involve a local cooling of 2 °C. These events may have had a significant influence on classical civilizations. For example, the cold arid event about 4,200 years ago coincides with the collapse of many classical civilizations, including the Old Kingdom in Egypt; the Akkadian Empire in Mesopotamia; the Early Bronze Age societies of Anatolia, Greece, and Israel; the Indus valley civilization in India; the Hilmand civilization in Afghanistan; and the Hongshan culture of China. The last of these millennial climate cycles was the Little Ice Age. This event is really two cold periods; the first follows the Medieval Warm Period, which ended 1,000 years ago, and is often referred to as the Medieval Cold Period. The Medieval Cold Period played a role in extinguishing Norse colonies on Greenland and caused famine and mass migration in Europe. It started gradually before AD 1200 and ended at about AD 1650. The second cold period, more classically referred to as the Little Ice Age, may have been the most rapid and greatest change in the North Atlantic region during the late Holocene, as suggested by ice-core and deep-sea sediment records. In Britain the temperature dropped on average by 1°C—although everyone assumes it was much greater because of the beautiful paintings of the ice fairs on the frozen River Thames. But this is a myth because it would be nearly impossible now to get the weather in England cold enough to freeze the River Thames, as it is no longer a slow meandering river due to the demolition of the old London Bridge in 1831, the straightening of the River Thames to build the Embankment in the 1870s so Londoners could promenade like the Parisians, and the dredging of the river to make it an international port at the heart of the British Empire.

If we examine records around the globe it seems that the Little Ice Age and the Medieval Warm Period only occurred in northern Europe, northeast America, and Greenland. So the Little Ice Age was a regional climate fluctuation driven by small changes in the Gulf Stream and deep-water formation north of Iceland. Many climate change deniers suggest that global warming is just the world recovering from the Little Ice Age. But as most of the world never had a Little Ice Age there is nothing to recover from. The reconstructed global temperature records for the last 2 millennia are essential as they provide a context for the instrumental temperature data set for the last 150 years. It's clearly shown that temperatures have been warmer in the 20th and the 21st centuries than at any other time during the last two thousand years.

Summary

For the last 2.5 million years the climate of the Earth has been dominated by the coming and going of the great ice sheets. These ice sheets were so thick that there was two miles of ice piled up on both North America and Northern Europe only 21,000 years ago. The changes in global climate were profound. During a glacial period the average global temperatures were 6 °C lower than today, global sea level was 120 m lower, and atmospheric carbon dioxide was reduced by a third and atmospheric methane by a half. The total weight of all the plants on the land was reduced by as much as half. The landscape of the planet was dramatically altered by the erosion and deposition of sediment by these huge ice sheets. Major rivers were re-routed and mountains cut in half. Land bridges appeared as the oceans lowered connecting continents, allowing species to colonize new lands. It also seems that over the last 2.5 million years the Earth's climate system would much prefer to be in a cold state rather than the warm one it is in today.

Chapter 8
Future climate change

Introduction

Future climate change is one of the defining challenges of the 21st century, along with poverty alleviation, environmental degradation, and global security. The problem is that 'climate change' is no longer just a scientific concern, it is now also of concern in terms of economics, sociology, geopolitics, national and local politics, law, and health, just to name a few. This chapter will examine briefly what 'anthropogenic' climate change is and the evidence that the global climate system is starting to change. The chapter will explain why changes in the climate system will lead to unpredictable weather patterns and an increase in the occurrence of extreme weather events, such as storms, floods, heatwaves, and droughts. Further details can be found in another of my titles in this series: *Global Warming: A Very Short Introduction*.

Human induced climate change

We have strong evidence that we have been changing the greenhouse gas content of the atmosphere. The first direct measurements of atmospheric carbon dioxide concentrations started in 1958 at an altitude of about 4,000 metres on the summit of Mauna Loa in Hawaii, a remote site free from local pollution. And looking even further back, air bubbles trapped in

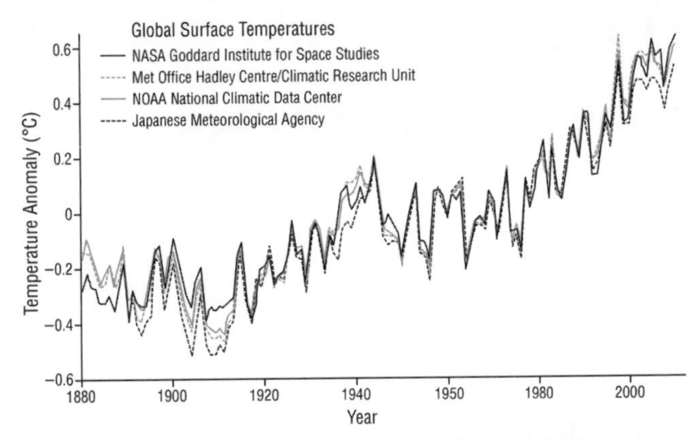

36. Global average surface temperature over the last 120 years

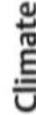

ice have been analysed from both the Greenland and the Antarctic ice sheets. These long ice core records suggest that pre-industrial carbon dioxide concentrations were about 280 parts per million by volume (ppmv). In 1958 the concentration was already 316 ppmv, and has climbed every year to reach 400 ppmv by June 2013. We have caused more pollution in one century than has occurred in thousands of centuries, throughout the natural waxing and waning of the great ice ages. Unfortunately, this increase in carbon dioxide in the atmosphere represents only half of the pollution we currently generate: about a quarter is absorbed by the oceans and another quarter by the land biosphere. One of the great worries scientists have is that this natural service may be reduced in the future making the situation worse.

According to the Intergovernmental Panel on Climate Change (IPCC) 2007 science report, increases in all greenhouse gases (see Chapter 2) over the last 150 years has already significantly changed the climate, including: an average rise in global temperatures of 0.75 °C; a rise in sea level of over 22 cm; a significant shift in the seasonality and intensity of precipitation, changing weather patterns; and a significant retreat of Arctic sea ice and nearly all continental glaciers. According to National Aeronautics and Space Administration (NASA), National Oceanic and Atmospheric Administration (NOAA), the UK Meteorological Office, and the Japanese Meteorological Agency, in the last 150 years the last decade has been the warmest on record (see Figure 36). The IPCC in 2007 stated that the evidence for climate change is unequivocal and there is very high confidence that this is due to human activity. This view is supported by a vast array of expert organizations, including the Royal Society and American Association for the Advancement of Science.

The 'weight of evidence'

Understanding future climate change is about understanding how science works and the principle of the 'weight of evidence'. Science moves forward by using detailed observation and experimentation to constantly test ideas and theories. Over the last 30 years the theory of climate change must have been one of the most comprehensively tested ideas in science. First, as described earlier, we have tracked the rise in greenhouse gases in the atmosphere. Second, we know from laboratory and atmospheric measurements that these gases do indeed absorb heat. Third, we have tracked significant changes in global temperatures and sea level rises over the last century. Fourth, we have analysed physical changes in the Earth's system related to climate, including the retreating sea ice around the Arctic and Antarctica, retreating mountain glaciers on all continents, and the shrinking of the area covered by permafrost with an increase in depth of its active layer. The ice cover records from the Tornio River in Finland, compiled since 1693, show that the spring thaw of the frozen river now occurs a month earlier. Fifth, we have tracked weather records and seen significant shifts. In recent years massive storms and subsequent floods have hit China, Italy, England, Korea, Bangladesh, Venezuela, Pakistan, Australia, and Mozambique. These observations are supported by detailed compilations of all precipitation records for the Northern

Hemisphere published in *Nature* in 2011 by Dr Seung-Ki Min and his colleagues in Canada showing a significant increase in the intensity of rainfall over the last 60 years. Moreover, in Britain the winter of 2000/1 was the wettest six months since records began in the 18th century; August 2008 was the wettest August on record; and April–June 2012 was the wettest spring on record. Also data collected by the British public show that birds are nesting 12±4 days earlier than 35 years previously. Sixth we have analysed the effects of natural changes on climate including sun spots and volcanic eruptions and though these are essential in understanding the pattern of temperature changes over the last 150 years they cannot explain the overall warming trend. And, last, we now understand longer term past climate changes and the role greenhouse gases have played in setting the climate of our planet.

'Climategate'

Despite all the evidence, discussion of future climate change evokes strong reactions. In part because many of the changes we might have to make to ameliorate the situation seem to go against the current neo-liberal market driven approach in the West. It is also due to a fundamental misunderstanding of science by the media, the public, and our politicians. This is beautifully discussed in Mark Henderson's book *The Geek Manifesto*. 'Climategate' and the other supposed climate change cover-ups reported in the media are excellent examples of this misunderstanding. Because science is not a belief system, you cannot decide you believe in antibiotics (as they may save your life) and metal tubes with sticky out bits can safely fly you across the Atlantic Ocean, but yet deny smoking can cause cancer, or that HIV causes AIDS, or that greenhouse gases cause global warming. This is because science is a self-correcting, rational methodology based on collecting and building up evidence, which is at the very foundation of our society. In the case of 'Climategate' there was in November 2009 an illegal release, due to hacking, of thousands of emails and other documents from the University of East Anglia's (UEA) Climatic Research Unit (CRU). Allegations were made that the emails revealed misconduct within the climate science community including the withholding of scientific information, prevention of papers being published, the deletion of raw data, and the manipulation of data to make the case for global warming appear stronger than it is. Three independent inquiries concluded that there was no evidence of scientific malpractice. But what all the media commentators missed at the time was that two other major groups at NOAA and NASA had used different raw data sets and different statistical approaches and published the very same conclusions as the UEA group. This was further supported in 2012 when Professor Richard Muller, a physicist and previously a climate change sceptic, and his Berkeley group published their collated global temperature records for the last 200 years and he publicly announced that he had changed his mind and that climate change *was* occurring due to human activity.

Figure 37 shows all of these composite data sets for global temperature over the last 2000 years, and not surprisingly they are different but they show very similar trends and all suggest that the 20th century was warmer than any other time in the last two millennia.

There was also the accusation that the UEA group and by extension other climate scientist had changed the raw data. Short hand terminology used by scientists such as 'correct', 'trick', 'tweak', 'manipulate', 'a line', and 'correlate' of course did not help this very much. However, some raw data does need to be processed so it can be compared with other data, particularly if you are trying to make long records of temperature and the methods used to measure temperature have changed during that time. The clearest example of this is the measurement of sea temperature, which up to 1941 was made in sea water hoisted on deck in a bucket. Originally these buckets were wooden then

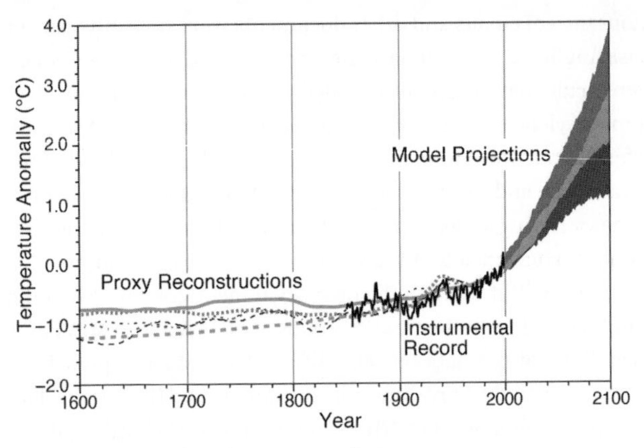

37. Past and future global average surface temperature

between 1856 and 1910 there was a shift to canvas buckets. This change in equipment would affect the level of cooling caused by evaporation as the water is being hoisted on deck. In addition, through this period there was a gradual shift from sailing ships to steamships, which altered the height of the ship decks and the speed of the ships, both of which could affect the evaporative cooling of the buckets. Since 1941, most sea temperature measurements have been made at the ships' engine water intakes—again, another shift. If scientists just stuck all this raw data together it would of course be wrong. Moreover in this case because the earlier sea-surface temperature measurements are too cold without correction it would make global warming in the ocean appear much greater than it really was. So the constant checking and correcting of data is extremely important in all parts of science. But the most important part is whether the results can be reproduced: is there a weight of evidence from many research groups showing the change? This is why after over 30 years of intensive research into climate change most scientists have a very high level of confidence that it is happening and it is due to human activity.

Climate change and its impact

In Chapter 3 we looked at how scientists model climate and future climate change. The synthesis of all the General Circulation Models is presented in the IPCC science reports. In 2007 they reported that global mean surface temperature could rise by between 1.1°C and 6.4°C by 2100, with best estimates being 1.8°C to 4°C. The biggest influence on how warm the climate becomes is that of emissions and what scenario is used. The faster and higher the greenhouse gas emissions rise, the hotter the world will become. It should be noted that global carbon dioxide emissions, despite the global recession, are rising as fast as the most dire 'business as usual' IPCC emission scenarios that have been looked at. The models also predict an increase in global mean sea level of between 18 cm and 59 cm. If the contribution from the melting of Greenland and Antarctica is included then this range increases to between 28 and 79 cm by 2100. All such predictions assume a continued linear response between global temperatures and ice sheet loss. This is unlikely, and sea level rise could thus be much higher. The next IPCC science report will be published in 2013 and will use more realistic future emission pathways, but draft chapters of this report show that they will reach very similar conclusions to the 2007 report.

The impacts of climate change will increase significantly as the temperature of the planet rises. The return period and severity of floods, droughts, heatwaves, and storms will increase. Coastal cities and towns will be especially vulnerable as sea levels rise, increasing the effects of floods and storm surges. A recent multi-disciplinary study by University College London published by the *Lancet* in 2009 demonstrated that the greatest threat of climate change to human health was from reduced water- and food-security, which could affect billions of people. Climate change also threatens the world's already devastated biodiversity. Ecosystems are already being hugely degraded by habitat loss, urbanization, pollution, and hunting. The 2007 'Millenium

Ecosystem Assessment' report suggested that three known species were becoming extinct each hour, while the 2008 Living Planet Index suggested that the global biodiversity of vertebrates had fallen by over a third in just 35 years, an extinction rate now 10,000 times faster than any observed in fossil records. The Royal Society's excellent 2012 'People and the Planet' report summarized the huge effects humans are having on the environment and how this will get worse as the global population increases but, more importantly, as consumption continues to rise uncontrollably around the world. Climate change of course will exacerbate all of this environmental degradation.

What is a 'safe' level of climate change?

So what level of climate change could be considered 'safe'? In February 2005 the British government convened an international science meeting in Exeter to discuss this very topic. This was so Tony Blair, the then prime minister, would have a political number to take to the G8 conference, which the UK was hosting later that year. Their recommendation was that global warming must be limited to a maximum of 2 °C above pre-industrial average temperature. Below this threshold it seems that there were both winners and losers due to regional climate change, but above this figure everyone seems to lose. This, however, is a purely political perspective, because if you live on any of the low lying Pacific Islands then your whole island may have been flooded by the time we reach this 2 °C level. However with the failure to produce a new climate treaty it now seems likely that temperature rises will exceed even this threshold. At the moment, in the 'business as usual' emissions scenario, we will hit that 2 °C point long before 2050. This is not surprising given that the International Energy Authority has predicted that the use of fossil fuels over the next 20 years will include a 30 per cent increase in the use of oil, a 50 per cent increase in the use of coal, and a 40 per cent increase in the use of natural gas.

So what is the cost of saving the world? According to the UK government-commissioned Stern Review on 'The Economics of Climate Change' in 2006, if we do everything we can now to reduce global greenhouse gas emissions and ensure we adapt to the coming effects of climate change, it will only cost us 1 per cent of world GDP every year. However, if we do nothing then the impacts of climate change could cost between 5 and 20 per cent of world GDP every year. These figures have been disputed. Some experts have argued that the cost of converting the global economy to lower carbon emissions could cost more than 1 per cent GDP because global emissions have risen faster than the worst predictions. In response Sir Nicolas Stern has recently revised his figure to 2 per cent of world GDP. Others argue that the costs could be offset by regional carbon trading systems. Others suggest that the impacts and the associated costs of global warming have been under-estimated by IPCC and the Stern Review. However, even if the cost–benefit of solving global warming is less than suggested by the Stern Review, there is an clear ethical case for preventing the deaths of tens of millions of people and the increase in human misery for billions that is otherwise to come.

Despite the complete failure to produce a new global climate change treaty there are countries and regions taking the reports seriously. The UK has introduced the long-term legally binding Climate Change Act. This Act provides a legal framework for ensuring that government meets the target of reducing greenhouse gas emissions by at least 80 per cent by 2050, compared to 1990 levels. In spring 2012 the UK was joined by Mexico who have ratified their own climate change national law which will cut their emissions by 50 per cent by 2050. In the European Union all member countries have agreed to the '20:20:20' policy by the year 2020. This is a bid to achieve a 20 per cent cut in greenhouse gas emissions, a 20 per cent increase in energy efficiency, and for 20 per cent of all energy to be produced from renewable sources.

Summary

By 2030 global food and energy demand will have increased by 50 per cent and water requirements will have increased by 30 per cent. This is partly due to the rise in global population but mostly it is caused by the rapid development of lower income countries. Added to this is the growing effects of climate change, which directly threaten water and food security, and you have what Sir John Beddington (UK Government Chief Scientific Adviser) calls the 'perfect storm' (Figure 38). Hence climate change and sustainable energy are the key scientific issues of the 21st century. We already have clear evidence for climate change with a 0.75°C rise in global temperatures and 22 cm rise in sea level during the 20th century. The IPCC predicts that global temperatures by 2100 could rise by between 1.8°C and 4.0°C, the range is due to the

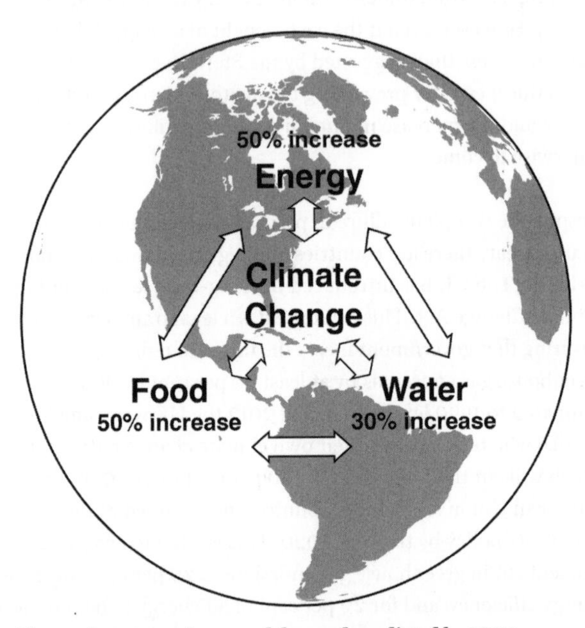

38. The perfect storm: increased demand predicted by 2030

uncertainty of how much greenhouse gas we will emit over the next 90 years. Sea levels could rise by between 28 cm and 79 cm—more, if the melting of the Greenland and Antarctica ice sheets accelerates. In addition, weather patterns will become less predictable and the occurrence of extreme climate events, such as storms, floods, heatwaves, and droughts, will grow more frequent. In the next chapter we will look at all the different options we have to fix our climate.

Chapter 9

Fixing climate change

Introduction

The most sensible approach to preventing the worst effects of future climate change would be to cut greenhouse gas emissions. Scientists believe a cut of between 50 per cent and 80 per cent in carbon dioxide by the middle of the century is required to avoid the worst effects. However, many have argued that the cost of significant cuts in fossil fuel use would severely affect the global economy. It would prevent rapid development and the alleviation of global poverty in a world where 8 million children die needlessly each year, 800 million people go to bed hungry each night, and 1,000 million people still do not have regular access to clean, safe drinking water. Nonetheless, under a business as usual emissions scenario there is a reasonable chance that we could be facing at least a 4°C warming by 2100, which would be disastrous and the impacts would fall disproportionally on the very poor in our society. In this chapter we will look at the three principal ways we could fix our climate. The first is 'mitigation' or reduction of the amount of carbon we emit. The second is the removal of carbon dioxide at source or from the atmosphere. The third is to use technology to reduce the amount of solar radiation being absorbed by the Earth, thus cooling the planet.

Mitigation

The idea of cutting global carbon emissions at least by half in the next 35 years and by up to 80 per cent by the end of the century may sound like fantasy, especially given the current trends in fossil fuel use (Figure 39). However, Steve Pacala and Robert Socolow, at Princeton University, published a very influential paper in the *Science* journal in a bid to make this challenge seem more achievable. They took the business as usual emissions scenario and the desired 450 parts per million by volume (ppmv) scenario and described the difference between the two as being represented by a number of 'wedges'. Instead of seeing one huge insurmountable problem, what we really have are lots of medium-sized changes, which would add up to the big change. They provided several examples for the wedges, each of them saving approximately 1 gigatonne of carbon every year. For example, one wedge would be doubling the efficiency of 2 billion cars from 30 miles per gallon (mpg) to 60 mpg, which actually is a very achievable aim, as there are family cars that have already been built that can easily do 100 mpg.

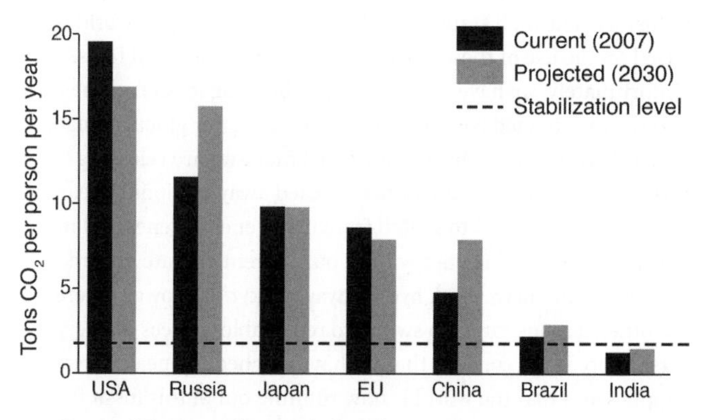

39. Past and future carbon emissions by country

One wedge is simply to improve energy efficiency. At the moment, the energy used in an average home in the USA is twice that of the average home in California: and the use in Californian domestic energy is twice that of the average home in Denmark. So already within the developed world there are huge savings to be made just by improving energy efficiency. If the industry and business sectors would reduce their energy use, they could significantly cut their running costs; however, this has not been an urgent priority for them due to the current incredibly low cost of energy. In any case, unfortunately it is likely that any efficiency gains these sectors made would ultimately be taken up by an increase in production and the level of energy use would increase accordingly. For example, if we did double the efficiency of 2 billion cars, this gain would be wiped out if production were increased by a further 2 billion cars. So, it is argued that the most important focus in cutting carbon dioxide production should be on the production of clean, or carbon-free, energy. This is discussed in the next section.

Alternative energy

The use of fossil fuels to provide energy has been an amazing discovery, allowing us to develop at a faster rate than at any time in history. The high standard of living in the developed world is based on the use of these cheap and relatively safe fossil fuels. Unfortunately, we have discovered that burning fossil fuels has had the unintended consequence of changing our global climate. This is because, as we burn these fossil fuels, we are releasing further solar energy that had been locked away millions of years ago. Instead, we need to switch from using energy generated in past climates to using energy from our current climate system. These include solar, wind, hydro-, wave, and tidal power. There are other reasons for us to switch to renewable sources of energy. First, there is the concern that we have reached the peak of our oil supplies and that the world is now running out of this fossil fuel. This may also be true for coal, despite there being hundreds of years' worth of high grade coal left. Second, countries have in the 21st century become very aware of 'energy security': most developed countries' economies are heavily reliant on imported fossil fuels, making them very vulnerable to volatility of the markets and international blackmail.

A brief discussion of the main alternative energy sources will now follow. It is immediately evident how certain sources of alternative energy would fit certain countries. For example, the UK has the best wind resource in the whole of Europe, while Saudi Arabia has excellent conditions for the sourcing of solar power. Almost all of the alternative sources use the climate system.

Solar energy

In Chapter 1 we discussed how the Earth receives on average 343 Watts per square metre (W/m^2) from the sun and that the Earth as a whole only received 2 billionths of all the energy put out by the sun. So the sun is in many ways the ultimate source of energy, which plants have been utilizing for billions of year. At the moment we can convert solar energy directly for heating or electricity, or we can capture the energy through photosynthesis by growing biofuels. The simplest approach is that of 'solar heating'. On a small scale, houses and other buildings in sunny countries could have solar heating panels on the roofs, which heat up water, for example, so people can have carbon-free hot showers and baths. On a large scale, parabolic mirrors can be used to focus solar energy to generate hot liquid (water or oil) with which to drive turbines to create electricity. The best places to situate solar heat plants are in low-latitude deserts, which have very few cloudy days per year. Solar heat plants have been built and used in California since the 1980s and are now being built and used in many other countries. Solar photovoltaic or solar panels convert sunlight directly into electricity. The individual rays of the sun hit the solar panel and dislodge electrons inside it, creating an electrical current. The main advantage of solar panels is that you can place them where the energy is needed and avoid the complicated infrastructure normally required to move electricity around. Over the last decade there has been a massive increase in their efficiency, the best commercially available solar panels are about 21 per cent efficient, which is great compared with photosynthesis that is about 1 per cent. There has been a significant drop in price in these solar panels with huge investment in the technology.

Biofuels

Biofuels are generated by the conversion of solar energy into plant biomass via photosynthesis, which can then be used to produce a liquid fuel. The global economy is based on the use of liquid fossil fuels, particularly for the transport sector. So in the short-term fuels derived from plants could be an intermediate low carbon way of powering cars, ships, and aeroplanes. There is the problem that the production of biofuels could compete with that of food crops. Indeed, the food price peaks of 2008 and 2011 were initially blamed on biofuel production. However, analysis by the New England Institute of Complexity showed these massive increases in price >50 per cent were in fact due to speculation on food prices in the financial markets.

Ultimately, electric cars are the future, because we can guarantee that the electricity produced is carbon-neutral. However, this is not an option for aeroplanes. Traditional aeroplane fuel is kerosene, which combines relative light weight with a high energy output. Research is being carried out to see whether a biofuel could be produced that would be light enough and powerful enough to replace kerosene.

Wind energy

Wind turbines are an efficient means of generating electricity—if they are large enough and preferably located out at sea. Ideally, you want turbines the size of the Statue of Liberty for maximum effectiveness. For example, the London Array is being built in the River Thames estuary and will generate 1,000 Megawatts (MW) of electricity, making it the world's largest consented offshore wind farm. When finished it could power up to 750,000 homes—about a quarter of Greater London—and reduce harmful carbon dioxide emissions by 1.4 million tonnes a year.

There are problems with wind turbines. First, they do not represent a constant supply of electricity: if the wind does not blow, then there is no electricity. Second, people do not like them, s they think they are ugly, noisy, and worry about the effects on local natural habitats. All these problems are easy to overcome by situating wind farms in remote locations, such as out at sea and away from areas of special scientific or natural interest. One study suggests that wind in principle could generate over 125,000 Terrawatt-hours, which is five times the current global electricity requirement.

Wave and tidal energy

Wave and tidal power could be an important source of energy in the future. The concept is simple, to convert the continuous movement of the ocean in the form of waves into electricity. However, this is easier said than done, and experts in the field suggest that wave power technology is now where solar panel technology was about 20 years ago—a lot of catching up is required. But tidal power in particular has one key advantage over solar and wind power: it is constant. In any country for the energy supply to be maintained there has to be at least 20 per cent production guaranteed, this is referred to as 'the baseline'. With the switch to alternative energy, new sources of power to ensure this constant baseline level need to be developed.

Hydroenergy

Hydroelectric power is an important source of energy globally. In 2010, it supplied 5 per cent of the world's energy. The majority of the electricity comes from large dam projects. These projects can present major ethical problems as large areas of land must be flooded above the dam, causing mass relocation of people and the destruction of the local environment. A dam also slows the flow of water in a river affecting the depositing of nutrient-rich silt in areas lower down on the river's course. If the river crosses national boundaries, there are potential issues over the rights to water and silt. For example, one of the reasons why Bangladesh is sinking is due to the lack of silt, and this lack has been caused by dams on the major rivers in India. There is also a debate about how much greenhouse gas hydroelectric plants save, because, even though the production of electricity does not cause any carbon emissions, the rotting vegetation in the area that has been flooded behind the dam does give off significant amounts of methane.

There are other sources of alternative or low-carbon energy which do not rely on the climate and these are briefly discussed in the next section, for completeness.

Geothermal energy

Below our feet, deep within the Earth, is hot, molten rock. In some locations, for example in Iceland and Kenya, this hot rock comes very close to the Earth's surface and can be used to heat water to make steam. This is an excellent, carbon-free source of energy, because the electricity generated from the steam can be used to pump the water down to the hot rocks. Unfortunately, this energy source is limited by geography. There is, however, another way this warmth of the Earth can be used. All new buildings could have a borehole below them with ground-sourced heat pumps. In this system, cold water is pumped down into these boreholes and the hot, molten rock below heats the water, cutting the cost of heating water, and this technology could be used almost everywhere in the world.

Nuclear fission

Energy is generated when you split heavy atoms such as uranium; this is known as nuclear fission. It has a very low direct carbon signature, however a significant amount of carbon is generated both from mining the uranium and from later decommissioning the power station. At the moment, 5 per cent of global energy comes from nuclear power. The new generation of nuclear power stations is extremely efficient: a level of nearly 90 per cent efficiency. The main disadvantages of nuclear power relate to the generation of high-level radioactive waste and the issue of safety. However, improvements in efficiency have reduced waste and the new generations of nuclear reactor sites have state-of-the-art safety precautions built in. The advantages of nuclear power are that it is reliable and can provide the required baseline in the energy mix; furthermore, it is a technology that is ready to go and has already been thoroughly tested.

Nuclear fusion

Nuclear fusion is the generation of energy when two smaller atoms are fused together. This is what happens in our sun and every other star. The idea is that the heavy form of hydrogen found in sea water can be combined and the only waste product is the non-radioactive gas helium. The problem, of course, is persuading those two atoms to join together. The sun does it by subjecting the atoms to incredibly high temperatures and pressures. Some advances have been made at the Joint European Torus (JET) project in the UK, which has produced 16 MW of fusion power. The problem is the amount of energy required to generate the huge temperatures in the first place and the difficulty of scaling the production up to power plant level.

Carbon dioxide removal

'Geoengineering' is the general term used for technologies that could be used to either remove greenhouses gases from the atmosphere or to change the climate of the Earth (Figure 40). Ideas considered under geoengineering range from the very sensible to the completely mad. At the moment we currently release over 8.5 gigatonnes of carbon (GtC) per year, so any interventions must be able to operate at a very large scale. In this section we will consider removal and disposal of atmospheric

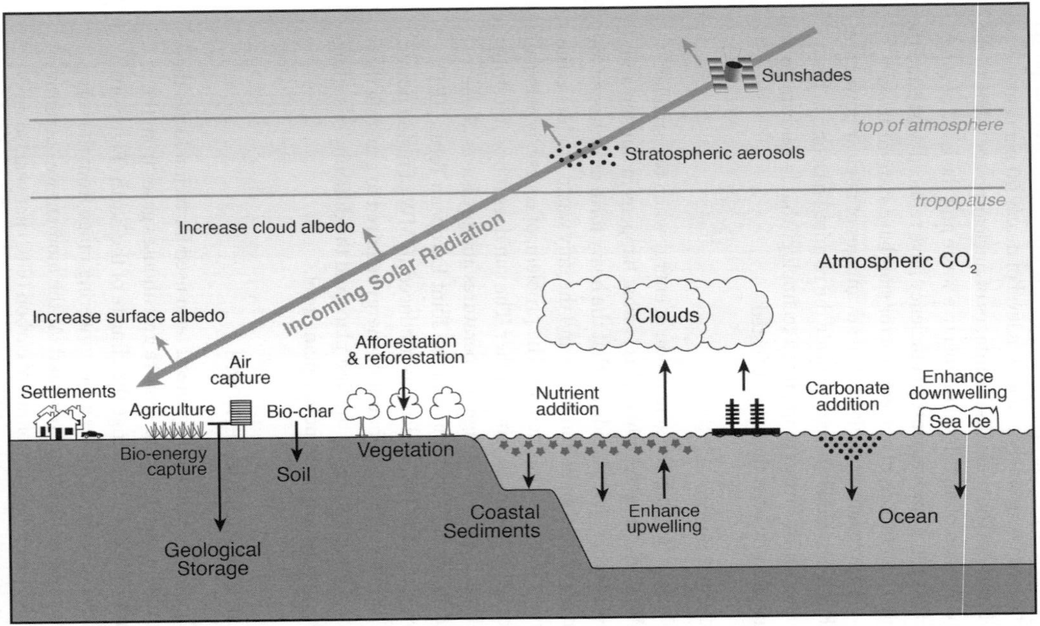

40. Geoengineering summary

carbon dioxide. There are three main approaches: biological, physical, and chemical.

Biological carbon removal

Biological approaches to carbon removal include the use of biofuels, discussed earlier, and reforestation. Reforestation or afforestation and avoidance of deforestation are sensible win-win solutions. By maintaining our forests we can retain biodiversity, stabilize soils and local rainfall, and provide livelihoods for local people via carbon credits. An excellent example of this can be seen in China. By 1990 the Loess Plateau, the bread-basket for China for at least the last 3,000 years, was turning into a dust bowl. Reforestation and overworking of the soils had started to reduce fertility, so farmers were cutting down more trees to open up more land to produce enough food to survive. The Chinese government became aware of this problem and over the subsequent 10 years instigated a radical tree-planting programme, with severe punishments for anyone caught chopping down trees. The effects were amazing, the trees stabilized the soils, greatly reducing soil erosion. The trees added moisture to the atmosphere, through transpiration, reducing evaporation and water loss. Once the forest reached a critical size and area it also started to stabilize the rainfall. In Chapter 8 we saw that the land's biosphere is already absorbing about 2 GtC per year of our pollution and Steve Pacala and Robert Socolow estimate that if we completely stopped global deforestation and doubled our current rate of planting we could produce another one of their 1 GtC per year wedges with all the win-win benefits that go with reforestation. In the UK the Forestry Commission proposed to increase the forested land from 12 per cent to 16 per cent by 2050. This would mean the government's target of 80 per cent reduction in carbon dioxide emissions by 2050 would only be 70 per cent, thanks to the absorption and storage of carbon by our forests.

The second biological approach is changing the uptake of carbon dioxide by the oceans. The most famous ocean 'technofix' was suggested by the late Professor John Martin. He suggested that many of the world's oceans are under-producing. This is because of the lack of vital micro-nutrients, the most important of which is iron, which allows plants to grow in the surface waters. Marine plants need minute quantities of iron and without it they cannot grow. In most oceans enough iron-rich dust gets blown in from the land, but it seems that large areas of the Pacific and Southern Ocean do not receive much dust and thus are barren of iron. So it has been suggested that we could fertilize the ocean with iron to stimulate marine productivity. The extra photosynthesis would convert more surface-water carbon dioxide into organic matter. Furthermore, when the organisms die, the organic matter drops to the bottom of the ocean, taking with it and storing the extra carbon. The reduced surface-water carbon dioxide is then replenished by carbon dioxide from the atmosphere. So, in short, fertilizing the world's oceans could help to remove atmospheric carbon dioxide and store it in deep-sea sediments. Experiments at sea have produced highly variable results, with some showing no effects at all while others revealing that the amount of iron required would be huge. Also, as soon as you stop adding the extra iron, most of this stored carbon dioxide is released, as very little organic matter is allowed to escape out of the photic zone (the part of the ocean close enough to the surface to receive the necessary sunlight for photosynthesis to take place) per year.

Physical

Removal of carbon dioxide during industrial processes is tricky and costly, because not only does the carbon dioxide need to be removed, but it must be stored safely. Removal and storage costs could be somewhere between $10 and $50 per tonne of carbon dioxide. This would cause a 15 per cent to 100 per cent increase in power production costs. However, due to the cheap and plentiful supply of high-grade coal, carbon capture and storage (CCS) is held to be one of the greatest hopes of governments worldwide. A lot more research is needed in this area to make CCS work efficiently and also to get it to an affordable level. The problem though is non-CCS coal or gas produced electricity will always be cheaper than CCS, so legislation is required to ensure companies are obliged to use CCS or to switch to alternative energy production. For example, the European Union (EU) emissions trading scheme (ETS), a 'cap-and-trade' system that all companies producing or using large amounts of energy are required to be part of, is helping reduce the EU total emissions by 2020.

Another possible solution is removal of carbon dioxide directly from the air. Considering carbon dioxide makes up just 0.04 per cent of the atmosphere this is much harder than it sounds. One mad idea is the production of artificial or plastic trees. Klaus Lackner a theoretical physicist and Allen Wright an engineer, supported by Wally Broecker, a climatologist, have designed carbon dioxide-binding plastic which can scrub carbon dioxide out of the atmosphere. In the proposed system, the carbon dioxide is then released from the plastic and taken away for storage. The first problem is water, as the plastic releases the carbon dioxide into solution when wet. This means that the plastic trees would have to be placed in very arid areas or would require giant umbrellas to protect them from rainfall. The second problem is the amount of energy that would be required to build, operate, and then store the carbon dioxide. The third problem is one of scale: tens of millions of these giant artificial trees would be required just to deal with US carbon emissions. When debating this approach on radio, I very gently suggested that perhaps we could just plant normal trees.

However, if plastic trees are not the answer, another form of technology to remove carbon dioxide either at source or ultimately from the atmosphere may be required.

Weathering

Carbon dioxide is naturally removed from the atmosphere over hundreds and thousands of years through the process of weathering. This process was discussed in Chapter 6 when we discussed the role of the uplift of the Himalayas in removing carbon dioxide from the atmosphere. Atmospheric carbon dioxide (CO_2) can interact with silica directly:

$$CaSiO_3 + CO_2 \rightarrow CaCO_3 + SiO_2$$

This is an extremely slow process removing less than 0.1 GtC per year, which is a hundred times less than we are emitting. Another process uses the combination of rainwater and carbon dioxide to form a weak carbonic acid solution:

$$CaSiO_3 + 2CO_2 + H2O \rightarrow Ca^{2+} + 2HCO_3^- + SiO_2$$

Only weathering of silicate minerals makes a difference to atmospheric carbon dioxide levels, as weathering of carbonate rocks by carbonic acid returns carbon dioxide to the atmosphere. By-products of hydrolysis reactions affecting silicate minerals are biocarbonates (HCO_3^-), metabolized by marine plankton and converted to calcium carbonate. The calcite skeletal remains of the marine biota are ultimately deposited as deep-sea sediments and hence lost from the global biogeochemical carbon cycle for the duration of the lifecycle of the oceanic crust on which they were deposited.

There are a number of geoengineering ideas aimed at enhancing these natural weathering reactions. One suggestion is to add silicate minerals to soils that are used for agriculture. This would remove atmospheric carbon dioxide and fix it as carbonate minerals and biocarbonate in solution. However the scale at which this would have to be done is very large and the effects on soils and their fertility are unknown. Another suggestion is to enhance the rate of reaction of carbon dioxide with basalts and olivine rocks in the Earth's crust. Concentrated carbon dioxide would be injected into the ground and would create carbonates deep underground. Again, like many geoengineering ideas, this is a great suggestion but very little work has been done to see if it is feasible, safe, and possible to do on the necessary scale.

Storage

Not all recovered carbon dioxide has to be stored; some may be utilized in enhanced oil recovery, the food industry, chemical manufacturing (producing soda ash, urea, and methanol), and the metal-processing industries. Carbon dioxide can also be applied to the production of construction material, solvents, cleaning compounds, packaging, and in waste-water treatment. But in reality, most of the carbon dioxide captured from industrial processes would have to be stored. It has been estimated that theoretically two-thirds of the carbon dioxide formed from the combustion of the world's total oil and gas reserves could be stored in the corresponding reservoirs. Other estimates indicate storage of 90–400 gigatonnes in natural gas fields alone and another 90 gigatonnes in aquifers.

Oceans could also be used to dispose of the carbon dioxide. Suggestions have included storage by hydrate dumping—if you mix carbon dioxide and water at high pressure and low temperatures, it creates a solid, or hydrate, which is heavier than the surrounding water and thus drops to the bottom. Another more recent suggestion is to inject the carbon dioxide half a mile deep into shattered volcanic rocks in between giant lava flows. The carbon dioxide will react with the water percolating through the rocks. The acidified water will dissolve metals in the rocks, mainly calcium and aluminium. Once it forms calcium bicarbonate with the calcium, it can no longer bubble out and escape. Though if it were to escape into the ocean, then bicarbonate would be relatively harmless. With ocean storage there is the added complication that we saw in Chapter 2: the ocean circulates, so whatever carbon dioxide you dump, some of it will eventually return. Moreover, scientists are very uncertain about the environmental effects this would have on the ocean ecosystems.

The major issue with all of these methods of storage is that of safety. Carbon dioxide is a very dangerous gas because it is heavier than air and can cause suffocation. An important example of this was a tremendous explosion of carbon dioxide from Lake Nyos that occurred in 1986, in the west of Cameroon, killing more than 1,700 people and livestock up to 25 km away. Though similar disasters had previously occurred, never had so many people and animals been asphyxiated on such a scale in a single brief event. What scientists now believe happened was that dissolved carbon dioxide from the nearby volcano seeped from springs beneath the lake and was trapped in deep water by the weight of the water above. In 1986, there was an avalanche that mixed up the lake waters, resulting in an explosive overturn of the whole lake, and all the trapped carbon dioxide was released in one go. However, huge amounts of mined ancient carbon dioxide are constantly being pumped around the USA to enhance oil recovery. There are no reports of any major incidents and engineers working on these pipelines feel they are much safer than they would be on gas or oil pipelines, which run across most major cities.

Solar radiation management

As we have seen, many of the geoengineering solutions proposed are still just ideas and need a lot more work to see if they are feasible. This is even more true of the solar radiation management ideas that have been proposed, many of which sound like something from a bad Hollywood B-movie. These suggestions include changing the albedo of the Earth: increasing the amount of solar energy reflected back into space to balance the heating from global warming (Figure 40). Methods to achieve this include: erecting massive mirrors in space; injecting aerosols into the atmosphere; making crops more reflective; painting all roofs white; increasing white cloud cover; and covering large areas of the world's deserts with reflective polyethylene-aluminium sheets. The fundamental problem with all of these approaches is that we have no idea what knock-on effects they would have. At the moment, we are performing one of the largest geoengineering experiments ever undertaken, by injecting huge amounts of greenhouse gases into the atmosphere, and though we have some idea of what might happen, we have no idea what the specific effects on our climate system could be. This is equally true of these geoengineering solutions—we currently have little idea if they would work or what unaccounted-for side-effects they might have. In many ways, climate change for the Earth can be seen in the same way as illness and the human body: it is always preferable to prevent an illness than to try and cure one, and we all know the potential side effects of drugs and chemo- or radiation therapy.

Let us use just one of these more far-fetched ideas as an example of what is wrong with solar radiation management: mirrors in space to deflect the sunlight. The most sophisticated of these suggestions is from Roger Angel, Director of the Centre for Astronomical Adaptive Optics at the University of Arizona, who suggests the use of a mesh of tiny reflectors to bend some of the light away from the Earth. He himself admits this would be expensive, requiring 16 trillion gossamer-light spacecraft costing at least $1 trillion and taking 30 years to launch. But, like all the other ideas we have discussed intended to change the Earth's albedo, it will not work. The reason for this is that all these approaches are fixated with getting the Earth's average temperature down, which misses out the importance of current temperature distribution, which as we know from Chapter 2 is what drives our climate. In fact, using climate models, Dan Lundt and colleagues at Bristol University show that these approaches would take us to a completely different global climate: the tropics would be 1.5 °C colder; the high latitudes would be 1.5 °C warmer; and precipitation would drop by 5 per cent globally compared with pre-industrial times.

Geoengineering governace

The Royal Society 2009 report on geoengineering not only reviewed the current scientific material in this area but also made the important step of trying to understand the governance issues associated with playing with the global climate system. There are a large number of ethical issues when considering how changing regional and global climate may affect countries differently. There may be overall positive results but minor changes in rainfall patterns could mean whole countries not receiving enough or too much rain, possibly resulting in disasters. The Royal Society summarized three main views on geoengineering:

1. It is a route to buy back some time to allow the failed international mitigation negotiations to catch up.
2. It represents a dangerous manipulation of the Earth's system and may be intrinsically unethical.
3. It is strictly an insurance policy against major mitigation policy failure.

Even if research is allowed to go ahead and geoengineering solutions are required, as with many emerging areas of modern technology, new flexible governance and regulatory frameworks will be required. Currently there are many international treaties which have a bearing on geoengineering and it seems that no single instrument applies. Hence 'fixing' our climate challenges our nation-state view of the world and new ways of governing will be required in the future.

Adaptation

Even if we decided to reduce carbon emissions significantly and tried all the geoengineering options available there would still be some climate change. This is because there has already been a temperature increase of 0.76°C, and even if we could reduce atmospheric carbon dioxide concentration to 2000 levels that would still add at least another 0.6°C. Considering the failure of international talks and the lack of serious investment in geoengineering we are currently on a 'business as usual' pathway. This means that many countries will be adversely affected by climate change in the near future, and nearly all countries will be affected in the next 30 years. So our failure to fix our climate issues means we must also plan to adapt to a new, changing climate. Each national government needs to research the vulnerabilities of their environment and socioeconomic system and predict the most likely effects of climate change for them.

The major threat from climate change, however, is its unpredictability. Humans can live in almost any extreme of climate from deserts to the Arctic, but only when we can predict the range of that extreme weather. So adaptation is really about how each country or region can deal with emerging threats of new levels of extreme weather. This adaptation should start now, as infrastructure changes, particularly in democracies, can take over 30 years to implement. For example, if you want to change land-use by building better sea defences or by changing farmland back to natural wetlands in a particular area, it can take up to 20 years to research and plan the appropriate measures. It can then take another 10 years for the full consultative and legal processes to be completed; a further 10 years to implement these changes; and another 10 years for the natural restoration to take place. A good example of this is the Thames Barrier, which currently protects London from flooding. It was built in response to the severe flooding in 1953, but it did not open officially until 1984—31 years later.

The other problem is that adaptation requires money to be invested now; and many countries just do not have the money. People do not want to pay more taxes to protect themselves in the future since most people live for today. Despite the fact that all of the adaptations discussed will in the long-term save money for the local area, the country, and the world, we as a global society still have a very short-term view—usually measured in the few years that elapse between successive governments.

The one thing that every government could do immediately is to set up a climate change impact assessment. For example, in the

UK there is the UK Climate Impact Programme (<http://www.ukcip.org.uk>), which shows the possible effects of climate change on the UK over the next 100 years. This programme is aimed at the UK national and local government, industry, business, the media, and the general public. If every government set up one of these programmes then at least their citizens would have the information to make informed choices about how their countries should be adapting to climate change.

Summary

So how should we fix global climate change? First it seems sensible to have an international political solution. We are currently without a post-2012 agreement and are looking at huge increases in global carbon emissions (Figure 41). Any political agreement will have to include measures to protect the rapid development of developing countries. It is a moral imperative that people in the poorest countries have the right to develop and to obtain the same lifestyle enjoyed by the developed world. We also need massive investment in alternative/renewable power sources and low carbon technology to provide the means of reducing world carbon emissions. There should be investment in geoengineering solutions, especially those that will make a significant impact in the short-term, such as reforestation and CCS. Action on climate change should also always contain an element of win-win. For example, supporting a huge increase in renewable energy not only reduces emissions but helps to provide energy security by reducing the reliance on imported oil, coal, and gas. Reduced deforestation and reforestation should not only draw down carbon dioxide from the atmosphere but will also help to retain biodiversity, stabilize soils, and provide livelihoods for local people via carbon credits. Measures that reduce car use will increase walking and cycling, which in turn should improve people's health, for example reducing levels of obesity and heart attacks.

We must not pin all our hopes on global politics, clean energy technology, and geoengineering—we must prepare for the worst

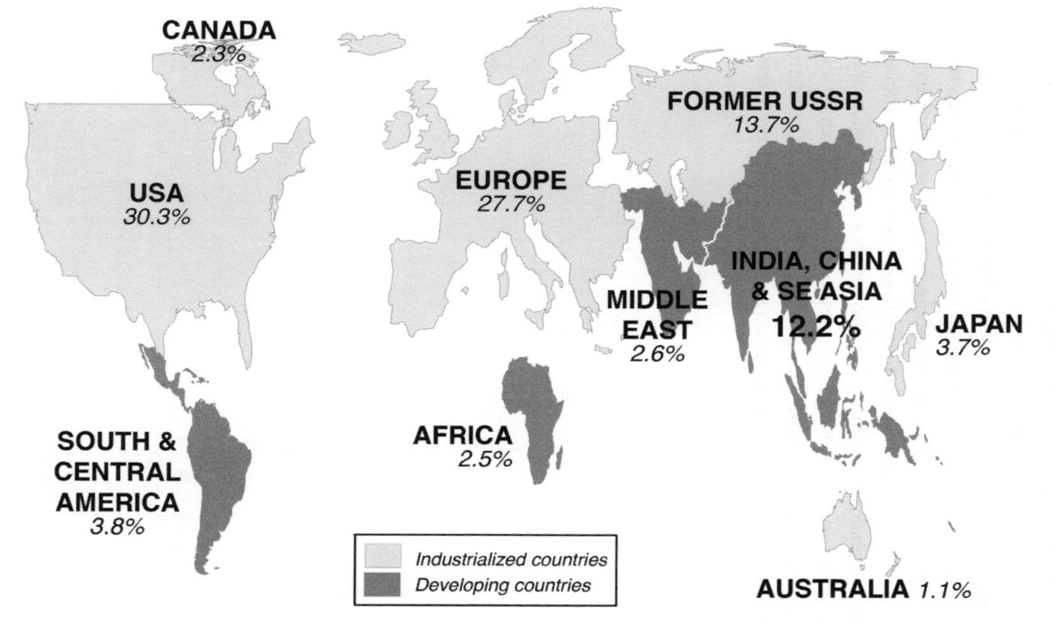

41. Historic proportion of carbon emissions

and adapt. If implemented now, much of the cost and damage that could be caused by changing climate can be mitigated. However, this requires nations and regions to plan for the next 50 years, something that most societies are unable to do due to the very short-term perspective of political institutions. This means that our climate issues are challenging the very way we organize our society. Not only do they challenge the concept of the nation-state versus global responsibility, but also the short-term vision of our political leaders. To answer the question of what we can do about climate change—we must change some of the basic rules of our society to allow us to adopt a much more global and long-term sustainable approach.

Chapter 10
Ultimate climate change

Introduction

By understanding how the climate of the Earth has varied in the past it is possible to look into the future. Many of the processes of change introduced in this book can be played forward to see what will happen in the near and far future.

The next ice age

The latest scientific research tells us the slow descent into a new ice age should start anytime in the next 1,000 years. As we saw in Chapter 7 we understand how the different wobbles of the Earth's orbit affect the Earth's climate. We can also look at previous interglacial periods to see how long they have lasted. Chronis Tzedakis, a professor of palaeoclimatology at University College London, and colleagues have calculated the natural length of the Holocene period and shown that any time from now until 1,500 years in the future the next ice age should start. But they also concluded that it probably will not happen. It seems that during every other interglacial period atmospheric carbon dioxide is highest as the climate system rebounds out of an ice age. Atmospheric carbon dioxide slowly lowers through each interglacial period until it hits a critical value of about 240 parts per million by volume (ppmv), which is 40 ppmv lower than pre-industrial times and 160 ppmv lower than today. Once this critical value has been reached then the climate system can respond to orbital forcing and slide into the next glacial period. However, it seems that our atmospheric carbon dioxide pollution, if it remains high, will prevent the world from sliding into an ice age over the next 1,500 years. In fact, according to the model predictions by Professor Andre Berger at the Université Catholique de Louvain in Belgium, if we double atmospheric carbon dioxide concentrations then global warming would postpone the next ice age for another 45,000 years. Interestingly, however, by that time the orbital forcing will be great enough to overcome extreme levels of greenhouse gases and normal glacial–interglacial cycles will reassert themselves.

Another interesting point is, why, even before the Industrial Revolution, was the atmospheric carbon dioxide level already above where we would have predicted it should be? This brings in the wonderful Ruddiman early anthropocene hypothesis. Bill Ruddiman, a professor of palaeoclimatology at the University of Virginia, has proposed that early agriculturalists caused a reversal in natural declines of atmospheric carbon dioxide starting about 7,000 years ago and of atmospheric methane starting about 5,000 years ago. This argument has caused huge controversy, but like all good theories it has been tested again and again, and no one has yet been able to disprove it. So essentially the argument is that early human interactions with the environment increased atmospheric greenhouse gases just enough that even prior to the Industrial Revolution there was enough of a change to the climate to delay the onset of the next ice age. It also raises the question of when exactly humans became such a geological force; and since all the ages of geology are given a name and there is a movement to define the age in which humans began to have a significant effect on the Earth's climate system.

The late ecologist Eugene F. Stoermer coined the term the 'Anthropocene', which has been popularized by the Nobel

Prize-winning atmospheric chemist Paul Crutzen. The term has Greek roots: *anthropo-* meaning 'human' and *-cene* meaning 'new'. Though the term is gaining a lot of support we still do not know where to put the boundary between the Holocene and the Anthropocene. In geology, boundaries between periods have to have a well defined datum or 'golden spike' in order to be recognized worldwide. Some have argued for Bill Ruddiman's proposed period of early human influence on atmospheric carbon dioxide, others have argued for the Industrial Revolution, and still others have argued for the period to be dated using trace element evidence of civilization (for example, the layer of chlorine from 1960s atomic weapon testing programmes) which can be found in ice cores. Whatever datum is finally chosen, there is no doubt that humans are now a major 'geological' force on the Earth: changing global patterns of erosion through massive land-use changes, including deforestation, which have caused mass extinction and huge loss of biodiversity, altering the global nitrogen cycle, ozone depletion, and of course climate change.

The next supercontinent

Christopher Scotese a professor at the University of Texas at Arlington is the director of the PALEOMAP Project. The project aims to illustrate how plate tectonics has changed ocean basins and continents and their positions over the last billion years. They also speculate how plate tectonics will change the face of the Earth in the future. In Chapter 5 we saw how supercontinents formed in the past and the severe effect they had on both climate and evolution. According to the PALEOMAP Project, the next supercontinent will form in the next 250 million years. Up to 50 million years in the future the world looks similar but with a few changes: the Atlantic Ocean will continue to widen; Africa will collide with Europe closing the Mediterranean Sea; Australia will collide with Southeast Asia; and California will slide northwards up the coast to Alaska (Figure 42). The really interesting changes happen between 50 and 150 million years in the future. Key to these are the major changes along the east coast of North and South America. Currently this is a passive margin and the continent and ocean plates are joined. But sometime around 50 million years in the future the Atlantic continents will pass over the Caribbean and Cocos plate boundary forming a new subduction zone (where tectonic plates collide, pushing one below the other) in the

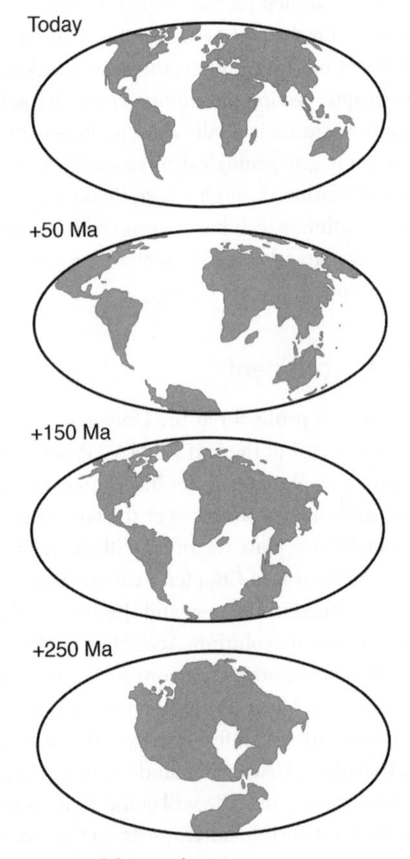

42. Future locations of the continents

Atlantic. This will create the Western Atlantic trench where the Atlantic Ocean continental plates will start to be consumed. Despite the mid-Atlantic ridge continuing to build new oceanic plate material, eventually the destruction will overtake the creation and the Atlantic Ocean will start to close up. At about 150 million years into the future the mid-Atlantic ridge will reach the subduction zone and be consumed. With no new ocean crust being created the closure of the Atlantic Ocean will then speed up. Elsewhere, the UK and Europe will then have a North Pacific Ocean view; the Mediterranean mountains will have reached their maximum height; and Antarctica and Australia will form one large continent. By 250 million years in the future the Americas, Africa, and Asia will have joined to form one supercontinent, with an inland sea the size of Australia. A small ocean gateway will separate this supercontinent from the Antarctic-Australian continent. We know from palaeoclimate records that supercontinents are bad for life. For example 96 per cent of all marine species and 70 per cent of terrestrial vertebrate species became extinct during the supercontinental Permian–Triassic extinction event 250 million years ago. So in another 250 million years, whatever life exists on Earth will face another major challenging period.

Boiling oceans

We assume the cycles of supercontinents forming and fragmenting will continue for as long as there are continents. However, there is a point where the climate will deteriorate to the point where complex, multi-cellular organisms cannot survive. For example, since the formation of our sun it has been increasing its energy output. The sun's luminosity increases about 10 per cent every billion years. Professor James Lovelock has suggested that feedbacks between life and climate system have modified the greenhouse gas content of the atmosphere to take account of this increase over the last few billion years; which is a central argument of his wonderful Gaia hypothesis. However in geological terms we already have extremely low levels of greenhouse gases, despite our best efforts over the last 100 years to reverse this. So the planet's ability to cool itself down is approaching its limit. This means that it is likely that over the next few billion years the temperature of the Earth will creep up (Figure 43). There will come a critical point when it is so warm that the oceans will start to evaporate, pumping huge amounts of moisture into the atmosphere. As we saw in Chapter 2 water vapour is one of the most important greenhouse gases and this runaway greenhouse would take average temperatures on Earth to over 100 °C, which no current multi-cellular organism can survive. I would suggest that ultimate climate change will occur when the oceans boil causing super global warming. It is rather ironic that the very place where life, and then complex life, evolved will be its ultimate destructor. Even the most extreme microbes will only last another 3 billion years and then it will be too hot even for them (Figure 43).

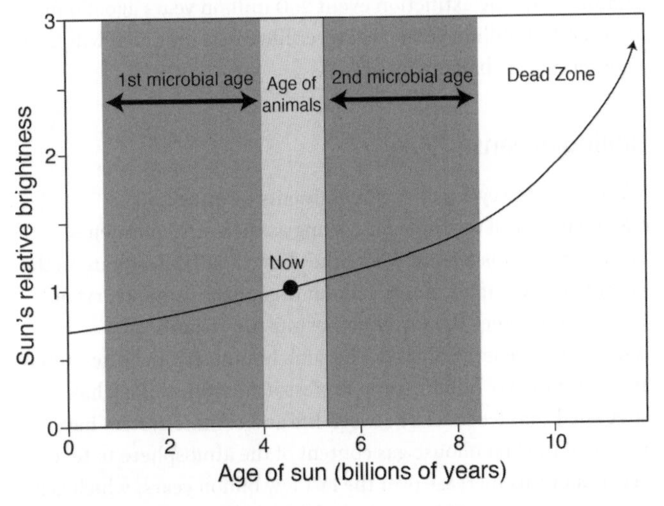

43. Life on Earth over the life of the sun

Death of the Earth

While multi-cellular life on Earth is likely to end in about a billion years' time and microbes in about 3 billion years' time, this would not be the end of the Earth. That is predicted to happen in about 5 billion years time. The sun is about halfway through turning its hydrogen into helium. Each second, more than 4 million tonnes of matter are converted into energy within the sun's core. Over the last 4.5 billion years the sun has turned 100 Earth-masses of matter into energy. The sun, however, does not have enough mass to explode as a supernova. Instead, in about 5 billion years it should enter a red giant phase. Its outer layers should expand as the hydrogen fuel at the core is consumed and the core should contract and heat up. Hydrogen fusion should then continue along a shell surrounding a helium core, which should steadily expand as more helium is produced. Once the core temperature reaches around 100 million °C the sun should even start producing carbon through the consumption of helium. As a red giant, the sun would be bigger than the Earth's current orbit area, about 250 times the present radius of the sun. It was thought that the Earth might survive even this, as the sun would have lost roughly 30 per cent of its present mass by then, so its gravity would be less strong, resulting in the orbits of the surrounding planets moving outwards. However, Peter Schröder of the University of Guanajuato in Mexico and Robert Smith of the University of Sussex in the UK created a detailed model of the sun's transition to a red giant and they found that the Earth's orbit would widen at first, but then the Earth would itself cause a 'tidal bulge' on the sun's surface from its own gravitational pull. The bulge would lag just behind the Earth in its orbit, slowing it down enough to drag it to a fiery demise. After the red giant phase and the death of the Earth, the sun would throw off its outer layers, forming a planetary nebula. The only object that would remain after the outer layers were ejected would be the extremely hot stellar core, which would slowly cool and fade over many billions of years to become a white dwarf star.

Death of the Earth

Further reading

Science of climate and weather

R. G. Barry and R. J. Chorley, *Atmosphere, Weather and Climate*, Routledge, 9th edition, p. 536 (2009)

A. Colling (ed.), *The Earth and Life: The Dynamic Earth*, Open University Worldwide, p. 256 (1997)

J. Gleick, *Chaos: Making a New Science*, Vintage, new edition, p. 380 (1997)

R. Hamblyn, *The Cloud Book: How to Understand the Skies*, David & Charles Publishers, p. 144 (2008)

T. R. Oke, *Boundary Layer Climates*, Routledge, second (reprinted) edition, p. 464 (2001)

Past climate

R. B. Alley, *The Two-Mile Time Machine: Ice Cores, Abrupt Climate Change, and Our Future*, Princeton University Press, new edition, p. 240 (2002)

B. Fagan (ed.), *The Complete Ice Age: How Climate Change Shaped the World*, Thames and Hudson, p. 240 (2009)

C. H. Langmuir and W. Broecker, *How to Build a Habitable Planet: The Story of Earth from the Big Bang to Humankind*, Princeton University Press, revised and expanded edition, p. 720 (2012)

J. J. Lowe and M. Walker, *Reconstructing Quaternary Environments*, Prentice Hall, 2nd edition, p. 472 (1997)

W. F. Ruddiman, *Earth's Climate: Past and Future*, W. H. Freeman, 2nd edition, p. 480 (2007)

R. C. L. Wilson, S. A. Drury, and J. L. Chapman, *The Great Ice Age: Climate Change and Life*, Routledge (2003)
J. Zalasiewicz and M. Williams, *The Goldilocks Planet: The 4 Billion Year Story of Earth's Climate*, Oxford University Press, p. 336 (2012)

Future climate change

A. Costello et al., 'Managing the Health Effects of Climate Change', *Lancet*, 373: 1693–733 (2009)
IPCC (Intergovernmental Panel on Climate Change), 'Climate Change 2007: The Physical Science Basis', Contribution of Working Group I to the Fourth Assessment Report of the Intergovernmental Panel on Climate Change, Solomon et al. (eds), Cambridge University Press (2007)
M. A. Maslin and S. Randalls (eds), *Future Climate Change: Critical Concepts in the Environment* (Routledge Major Work Collection: 4 volumes containing reproductions of 85 of the most important papers published in *Climate Change*) p. 1600 (2012)
Mark Maslin, *Global Warming: A Very Short Introduction*, Oxford University Press, second edition, p. 192 (2008)
W. F. Ruddiman, *Plows, Plagues, and Petroleum: How Humans Took Control of Climate*, Princeton University Press, new edition, p. 240 (2010)
N. Stern, *The Economics of Climate Change: The Stern Review*, Cambridge University Press, p. 692 (2007)
G. Walker and D. King, *The Hot Topic*, Bloomsbury, p. 309 (2008)

Fixing climate

R. Gelbspan, *Boiling Point*, Basic Books, p. 254 (2005)
Mark Henderson, *The Geek Manifesto: Why Science Matters*, Bantam Press (2012)
M. Hillman, *How We Can Save the Planet*, Penguin Books (2004)
R. Kunzig and W. Broecker, *Fixing Climate*, GreenProfile, in association with Sort of Books, p. 288 (2008)
C. Hamilton, *Earthmasters: The Dawn of the Age of Climate Engineering*, Yale University Press, p. 247 (2013)
M. A. Maslin and J. Scott, 'Carbon Trading Needs a Multi-Level Approach?' *Nature*, 475: 445–7 (2011)

A. Meyer, *Contraction and Convergence: The Global Solution to Climate Change*, Green Books (2000)

The Royal Society, 'Geoengineering the Climate: Science, Governance and Uncertainty', The Royal Society Science Policy Centre report 10/09, The Royal Society, p. 81 (2009)

General reading

D. Brownlee and P. Ward, *The Life and Death of Planet Earth: How Science Can Predict the Ultimate Fate of Our World*, Piatkus, p. 256 (2007)

J. D. Cox, *Weather for Dummies*, first edition, p. 384 (2000)

R. Hamblyn, *Extraordinary Weather: Wonders of the Atmosphere from Dust Storms to Lightning Strikes*, David & Charles Publishers, p. 144 (2012)

J. Martin, *The Meaning of the 21st Century*, Eden Project Books, p. 526 (2007)

The Royal Society, 'People and the Planet', The Royal Society Science Policy Centre report 01/12, The Royal Society, p. 81 (2012)

Fiction inspired by climate

D. Defoe, *The Storm*, Penguin Classics, new edition, p. 272 (2005)

K. Evans, *Funny Weather*, Myriad Editions, p. 95 (2006)

J. Griffiths, *WILD: An Elemental Journey*, Penguin Books (2008)

P. F. Hamilton, *Mindstar Rising*, Pan Books (1993)

S. Junger, *The Perfect Storm: A True Story of Man Against the Sea*, Harper Perennial, reissue edition, p. 240 (2006)

J. McNeil, *The Ice Lovers: A Novel*, McArthur & Company, p. 325 (2009)

K. S. Robinson, *Forty Signs of Rain*, HarperCollins (2004)

Index

D

E

F

G

H

I

J

K

L

M

N

O

P

R

S

T

Expand your collection of
VERY SHORT INTRODUCTIONS

1. Classics
2. Music
3. Buddhism
4. Literary Theory
5. Hinduism
6. Psychology
7. Islam
8. Politics
9. Theology
10. Archaeology
11. Judaism
12. Sociology
13. The Koran
14. The Bible
15. Social and Cultural Anthropology
16. History
17. Roman Britain
18. The Anglo-Saxon Age
19. Medieval Britain
20. The Tudors
21. Stuart Britain
22. Eighteenth-Century Britain
23. Nineteenth-Century Britain
24. Twentieth-Century Britain
25. Heidegger
26. Ancient Philosophy
27. Socrates
28. Marx
29. Logic
30. Descartes
31. Machiavelli
32. Aristotle
33. Hume
34. Nietzsche
35. Darwin
36. The European Union
37. Gandhi
38. Augustine
39. Intelligence
40. Jung
41. Buddha
42. Paul
43. Continental Philosophy
44. Galileo
45. Freud
46. Wittgenstein
47. Indian Philosophy
48. Rousseau
49. Hegel
50. Kant
51. Cosmology
52. Drugs
53. Russian Literature
54. The French Revolution
55. Philosophy
56. Barthes
57. Animal Rights
58. Kierkegaard
59. Russell
60. Shakespeare
61. Clausewitz
62. Schopenhauer
63. The Russian Revolution
64. Hobbes
65. World Music

66. Mathematics
67. Philosophy of Science
68. Cryptography
69. Quantum Theory
70. Spinoza
71. Choice Theory
72. Architecture
73. Poststructuralism
74. Postmodernism
75. Democracy
76. Empire
77. Fascism
78. Terrorism
79. Plato
80. Ethics
81. Emotion
82. Northern Ireland
83. Art Theory
84. Locke
85. Modern Ireland
86. Globalization
87. The Cold War
88. The History of Astronomy
89. Schizophrenia
90. The Earth
91. Engels
92. British Politics
93. Linguistics
94. The Celts
95. Ideology
96. Prehistory
97. Political Philosophy
98. Postcolonialism
99. Atheism
100. Evolution
101. Molecules
102. Art History
103. Presocratic Philosophy
104. The Elements
105. Dada and Surrealism
106. Egyptian Myth
107. Christian Art
108. Capitalism
109. Particle Physics
110. Free Will
111. Myth
112. Ancient Egypt
113. Hieroglyphs
114. Medical Ethics
115. Kafka
116. Anarchism
117. Ancient Warfare
118. Global Warming
119. Christianity
120. Modern Art
121. Consciousness
122. Foucault
123. The Spanish Civil War
124. The Marquis de Sade
125. Habermas
126. Socialism
127. Dreaming
128. Dinosaurs
129. Renaissance Art
130. Buddhist Ethics
131. Tragedy
132. Sikhism
133. The History of Time
134. Nationalism
135. The World Trade Organization
136. Design
137. The Vikings
138. Fossils
139. Journalism

140. The Crusades
141. Feminism
142. Human Evolution
143. The Dead Sea Scrolls
144. The Brain
145. Global Catastrophes
146. Contemporary Art
147. Philosophy of Law
148. The Renaissance
149. Anglicanism
150. The Roman Empire
151. Photography
152. Psychiatry
153. Existentialism
154. The First World War
155. Fundamentalism
156. Economics
157. International Migration
158. Newton
159. Chaos
160. African History
161. Racism
162. Kabbalah
163. Human Rights
164. International Relations
165. The American Presidency
166. The Great Depression and The New Deal
...7. Classical Mythology
... The New Testament as ...terature
...rican Political Parties
...ctions
...s

175. Documentary Film
176. Modern China
177. The Quakers
178. German Literature
179. Nuclear Weapons
180. Law
181. The Old Testament
182. Galaxies
183. Mormonism
184. Religion in America
185. Geography
186. The Meaning of Life
187. Sexuality
188. Nelson Mandela
189. Science and Religion
190. Relativity
191. The History of Medicine
192. Citizenship
193. The History of Life
194. Memory
195. Autism
196. Statistics
197. Scotland
198. Catholicism
199. The United Nations
200. Free Speech
201. The Apocryphal Gospels
202. Modern Japan
203. Lincoln
204. Superconductivity
205. Nothing
206. Biography
207. The Soviet Union
208. Writing and Script
209. Communism
210. Fashion
211. Forensic Science
212. Puritanism

213. The Reformation
214. Thomas Aquinas
215. Deserts
216. The Norman Conquest
217. Biblical Archaeology
218. The Reagan Revolution
219. The Book of Mormon
220. Islamic History
221. Privacy
222. Neoliberalism
223. Progressivism
224. Epidemiology
225. Information
226. The Laws of Thermodynamics
227. Innovation
228. Witchcraft
229. The New Testament
230. French Literature
231. Film Music
232. Druids
233. German Philosophy
234. Advertising
235. Forensic Psychology
236. Modernism
237. Leadership
238. Christian Ethics
239. Tocqueville
240. Landscapes and Geomorphology
241. Spanish Literature
242. Diplomacy
243. North American Indians
244. The U.S. Congress
245. Romanticism
246. Utopianism
247. The Blues
248. Keynes
249. English Literature
250. Agnosticism
251. Aristocracy
252. Martin Luther
253. Michael Faraday
254. Planets
255. Pentecostalism
256. Humanism
257. Folk Music
258. Late Antiquity
259. Genius
260. Numbers
261. Muhammad
262. Beauty
263. Critical Theory
264. Organizations
265. Early Music
266. The Scientific Revolution
267. Cancer
268. Nuclear Power
269. Paganism
270. Risk
271. Science Fiction
272. Herodotus
273. Conscience
274. American Immigration
275. Jesus
276. Viruses
277. Protestantism
278. Derrida
279. Madness
280. Developmental Biology
281. Dictionaries
282. Global Economic History
283. Multiculturalism
284. Environmental Economics
285. The Cell
286. Ancient Greece
287. Angels

288. Children's Literature
289. The Periodic Table
290. Modern France
291. Reality
292. The Computer
293. The Animal Kingdom
294. Colonial Latin American Literature
295. Sleep
296. The Aztecs
297. The Cultural Revolution
298. Modern Latin American Literature
299. Magic
300. Film
301. The Conquistadors
302. Chinese Literature
303. Stem Cells
304. Italian Literature
305. The History of Mathematics
306. The U.S. Supreme Court
307. Plague
308. Russian History
309. Engineering
310. Probability
311. Rivers
312. Plants
3. Anaesthesia
The Mongols
he Devil
ectivity
etism
322. Stars
323. The Antarctic
324. Radioactivity
325. Trust
326. Metaphysics
327. The Roman Republic
328. Borders
329. The Gothic
330. Robotics
331. Civil Engineering
332. The Orchestra
333. Governance
334. American History
335. Networks
336. Spirituality
337. Work
338. Martyrdom
339. Colonial America
340. Rastafari
341. Comedy
342. The Avant-Garde
343. Thought
344. The Napoleonic Wars
345. Medical Law
346. Rhetoric
347. Education
348. Mao
349. The British Constitution
350. American Politics
351. The Silk Road
352. Bacteria
353. Symmetry
354. Marine Biology
355. The British Empire
356. The Trojan War
357. Malthus
358. Climate